Contents

Content Guidance

Getting the most from this book

Examiner tips

Advice from the examiner on key points in the text to help you learn and recall unit content, avoid pitfalls, and polish your exam technique in order to boost your grade.

Knowledge check

Rapid-fire questions throughout the Content Guidance section to check your understanding.

Knowledge check answers

1 Turn to the back of the book for the Knowledge check answers.

Summary

Summaries

● Each core topic is rounded off by a bullet-list summary for quick-check reference of what you need to know.

Content Guidance

Exam-style questions →

Sample student answers
Practise the questions, then look at the student answers that follow each set of questions.

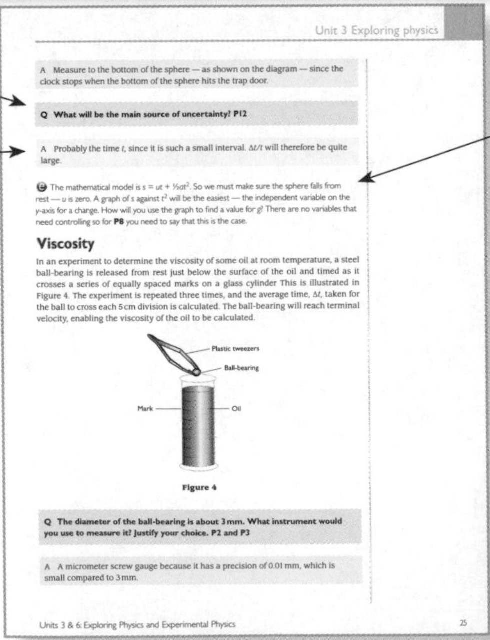

Unit 3 Exploring physics

A Measure to the bottom of the sphere — as shown on the diagram — since the clock stops when the bottom of the sphere hits the trap door.

Q What will be the main source of uncertainty? P12

A Probably the time t, since it is such a small interval. $\Delta t/t$ will therefore be quite large.

ⓔ The mathematical model is $s = ut + \frac{1}{2}at^2$. So we must make sure the sphere falls from rest — u is zero. A graph of s against t^2 will be the easiest — the independent variable on the y-axis for a change. How will you use the graph to find a value for g? There are no variables that need controlling so for **P8** you need to say that this is the case.

Viscosity

In an experiment to determine the viscosity of some oil at room temperature, a steel ball-bearing is released from rest just below the surface of the oil and timed as it crosses a series of equally spaced marks on a glass cylinder This is illustrated in Figure 4. The experiment is repeated three times, and the average time, Δt, taken for the ball to cross each 5 cm division is calculated. The ball-bearing will reach terminal velocity, enabling the viscosity of the oil to be calculated.

Plastic tweezers
Ball-bearing
Mark — Oil

Figure 4

Q The diameter of the ball-bearing is about 3 mm. What instrument would you use to measure it? Justify your choice. P2 and P3

A A micrometer screw gauge because it has a precision of 0.01 mm, which is small compared to 3 mm.

Units 3 & 6: Exploring Physics and Experimental Physics   25

Examiner commentary on sample student answers
Find out how many marks each answer would be awarded in the exam and then read the examiner comments (preceded by the icon ⓔ) following each student answer.

About this book

Physics is a practical science, which means that you will do a lot of experiments! When a tree is chopped down it falls over and ends up lying on the ground. You know about Isaac Newton watching an apple fall, so, if asked you will probably say that the tree falls because of gravity. You might then say that gravity attracts the tree and go on to say that the mass of the Earth attracts the mass of the tree and that when the tree is free to move it gets closer to the Earth. Watching a diving competition you might notice that all the divers take the same time to get to the water — it doesn't matter how much gravity attracts each one. Galileo was one of the first physicists to explain this. Physicists will only give such an explanation after carrying out lots of experiments and making lots of observations on many different things falling — not just trees and divers. The whole of physics starts with experiments.

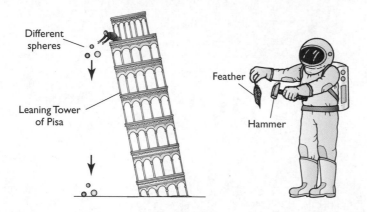

As part of your study of physics you can expect to do a number of experiments to see if the explanations are true — was Galileo right? Maths has a part to play in the development of mathematical models — equations — that can be tested by taking measurements. If the model predicts correctly what happens next time, then it is a good model. However, you won't know this until the model has been tested by experiment. This means looking at things, changing something and then looking again. You need to know what you are aiming for and you have to plan what to do. When you have finished an experiment you have to decide whether you have found out anything useful.

Part of your A-level assessment is practical work and you will do some coursework at both AS and A2. Unit 3 is the AS coursework; Unit 6 is the A2 coursework. Each of these units counts for 10% of the total marks. This guide is part of a series covering the whole of the Edexcel specification. The aim is to explain how your practical work is assessed and to help you gain as many marks as you can. In order to build up your skills you will need to do a lot of physics practicals. If you keep this guide with you as you carry out the work you will be able to make sure you are doing it correctly.

For each unit there is an overview, followed by the assessment criteria and advice on carrying out practical work. Each assessment criterion has a skill code — for example **P8** and **A14** — and when these occur they are in bold to provide you with a signpost, so look out for these. There are also some worked examples and an exemplar script

for each unit. The exemplar is a piece of work produced by a student. The way it has been marked is described and some explanations of how improvements could be made are given, preceded by the (e) icon. Throughout the text there are Knowledge check questions for you to work through. Answers are provided at the end of the book.

Planning, carrying out and analysing your own work are difficult skills. You will, however, get better with practice. This is a guide to that practice.

Content Guidance

Unit 3 Exploring physics

Overview

The practical assessment for AS is based firmly in the real world of physics applications. You will start by going on a visit or doing a case study into something you would like to know more about. This unit is a bridge between the skills you needed for GCSE and those that will take you up to A2. It is important to remember and build on your existing skills and knowledge and at the same time discover new areas for investigation and be able to recognise where the demand is greater. This unit is assessed by means of 40 assessment criteria, each of which consists of a simple statement. You will have a copy of the criteria as you carry out the work, so it is a good idea to first familiarise yourself with what the criteria say — we shall do this in the next section. Each criterion is worth 1 mark, so the total mark for Unit 3 is 40. Each criterion stands alone, i.e. you do not have to satisfy one of the criteria before you go on to another.

There are two parts to this unit, the report and the practical work. The two parts are linked and you will be given a **briefing sheet** that provides an overall direction; it is important that you read this carefully before you start. It is the report that you write on either a visit or a case study that is marked. This is then followed by three phases of practical work:
- planning
- implementation and measurements
- analysis

The work is chosen by your school or college and will probably be on one of the topics you have been studying — but it does not have to be. Since you need to develop your practical skills this unit will probably be taken in the second term of your course.

The report on the visit or case study is written in your own time, and you may use ICT to produce a good-looking finished product. You will do the practical work during class time and there are no time constraints so everything is done in either your physics laboratory or classroom. You work on your own and you must hand in your work each time you leave.

Report

The report is worth a total of 11 marks: 9 marks for the content and 2 marks for the structure of the report and the standard of written English.

You write a report of no more than 600 words on either a visit or a case study. It will be marked for your use of English so you must write it as well as you can — ICT will help here. After this there will be some practical work that is linked to your visit or case study. You will be given a **briefing sheet** that will introduce your specific work.

Visit

This can be almost anywhere — a theme park, a concert hall, an industrial installation such as a manufacturing research site or even a power station. The point is that it is a physics visit and you will be looking at an application of physics, but you can certainly have a good time while you are there. It would be a good idea to visit the relevant website before you go on the visit so that you know what questions to ask.

Case study

In effect, this is a small research project. You consult a variety of sources and write a report on what you find out. The advantage of doing a case study is that you can study some new ideas or perhaps some applications that require facilities to which you do not have access. The case study is carried out entirely in your own time and there is no travelling involved.

Planning

The planning section is worth 14 marks.

The **briefing sheet** provides information about what you will be investigating. It will set the work in the context of your case study or visit and provide you with an aim. The work should not come as a surprise because your teacher will most likely have told you what you will be doing during the practical, but not the actual aim of the experiment.

Planning your practical work starts after the visit or case study is complete. You might find that the work you did for the report will help in your plan. From now on all the work you do is done on your own, supervised by a teacher. You must hand in all your work each time you leave the laboratory or classroom. When you write your plan you will have the list of assessment criteria with you. This will help you to keep in line with what you have to do to score marks. Writing the plan should take no more than about an hour.

Implementation and measurements

The implementation and measurements section is worth 4 marks.

Now you carry out your plan. This part is really just ordinary laboratory work (such as that which you have done during your time studying physics) and will probably take less than an hour. You will need to think about how to make your results as accurate and reliable as possible. You should record your results in a table, with units

Edexcel AS/A2 Physics

Analysis

The analysis section is worth 11 marks.

The variables in your experiment will have a fairly simple relationship. You will need to plot a graph of your results so that you have a straight line (probably) and then find its gradient. From this you will go on to see how well you have met the aim. You should also think about the uncertainties in your work and finish with a conclusion that relates to the aim on the briefing sheet.

The analysis should take about an hour.

After writing the report on the visit or case study you do the following:

- **(1)** Read the briefing sheet again.
- **(2)** Plan the practical work.
- **(3)** Carry out the practical work.
- **(4)** Analyse the results.
- **(5)** Write a conclusion and evaluate your work.

Summary

Assessment criteria

You will find a copy of the assessment criteria most helpful. A complete list is available from the Edexcel website at: www.edexcel.com/quals/gce/gce08/physics/Pages/default.aspx

If you click on 'Assessment Materials' you will find a document called 'GCE Physics 2008 marking grids AS'; this contains an exact statement of all the criteria. There is 1 mark for each of the criteria and they are all independent. Since the criteria are short, what follows is an indication of what you should do in order to score the mark in each case.

Summary of case study or physics-based visit

Ref	What you must do to score the mark	
	Visit	**Case study**
S1	Make a visit away from school to find out about an application of physics and produce a report that addresses the criteria listed below.	Write a report summarising some research you have carried out. You must consult a minimum of three different sources of information. Note that, for example, three websites count as one type of source so you must also look at other sources, such as books, journals, magazines and manufacturers' data sheets.
S2	Give details of where you went to visit and the type of institution it is.	You must provide full details of sources of information. This includes full web addresses with the dates visited, authors and publishers of books or magazines and so on.
S3	What is the application of physics that you are looking at in your visit?	What is the application of physics that you are looking at in your case study?
S4	You should describe how physics principles are relevant to your study.	
S5	Make sure that you use relevant specialist terminology correctly.	
S6	You should include in your report one piece of relevant information (e.g. data, graph, diagram) that you found out for yourself and which is not mentioned in the briefing sheet.	
S7	Describe how the physics principles in **S4** fit into the context of your study. This can be any context — social, environmental, historical or even health and safety.	
S8	Describe the implication of the physics. How does the use of physics change what is going on; how does physics bring benefits to people or reduce risks?	
S9	You should explain how the practical work relates to the physics in your report.	
	Maximum marks for this section — 9	

Examiner tip
Read the criteria carefully. If a criterion says comment, then a simple statement in your report will not get the mark. For example, look at **P9**. Simply stating that you will take repeats is not enough, you must discuss why doing this will improve your results.

Report

Ref	What you must do to score the mark
R1	Your summary report should be easy to follow and contain few grammatical or spelling errors.
R2	Your summary report should be structured and use appropriate subheadings.
	Maximum marks for this section — 2 **These 2 marks are only for the report on the visit or case study.**

Planning

Ref	What you must do to score the mark
P1	Make a list of all the materials required. It is a good idea to draw a diagram (**P13**) at this stage and show on it how your apparatus will work together.
P2	State clearly how you will measure one relevant quantity using the most appropriate instrument.
P3	Use precision and uncertainty to explain your choice of measuring instrument, possibly including the number of measurements to be taken.
P4	As for **P2**, state clearly how you will measure another relevant quantity using the most appropriate instrument. This can be any of the measurements you intend to make, not necessarily one of the variables.
P5	As for **P3**, use precision and uncertainty to explain your choice of measuring instrument, possibly including the number of measurements to be taken. This might be using the same instrument to take a different measurement.
P6	Describe any technique that demonstrates what you might do to make your measurement accurate and reliable — for example, a zero-error check or stirring a liquid. Say why your technique improves the measurement.
P7	State which is the independent variable and which is the dependent variable.
P8	Make sure that your experiment will be a fair test by identifying other relevant quantities and either explain how you will control these or state that, as they will not change significantly, they will not affect the results.
P9	Explain how repeat readings would improve your results and whether repeat readings are appropriate in your experiment.
P10	If the equipment is safe you should explain why. Otherwise, identify the hazards and give simple precautions.
P11	Explain exactly how you will use the data collected to obtain the value mentioned in the briefing sheet.
P12	Think about the readings or methods that are the main sources of uncertainty and/or systematic error and write a brief description.
P13	Use a ruler and a sharp pencil to draw an appropriately labelled large diagram of the apparatus to be used.
P14	Make sure your plan is well organised and methodical; a bullet point sequence works well.
	Maximum marks for this section — 14

Implementation and measurements

Ref	What you must do to score the mark
M1	Record all your measurements, including repeats, in a table drawn before you start. Use the correct number of significant figures as determined by the precision quoted in **P3** and **P5**.
M2	Make sure the units are correct throughout the whole practical report.
M3	Obtain enough measurements to plot a graph, i.e. six for a straight-line graph and possibly more for a curve.
M4	Obtain measurements to give a reasonable range; this might be so that one of the variables doubles or halves.
	Maximum marks for this section — 4

Analysis

Ref	What you must do to score the mark
A1	Draw a graph with appropriately labelled axes and with correct units.
A2	The scales for the graph should be easy to read and the points should occupy as much of the paper as possible.
A3	Points should be accurate to ±1 mm; this is usually half a small square.
A4	The line of best fit can be either a straight line or a smooth curve. It should not be forced through the origin and should show the trend.
A5	Comment on the trend or pattern obtained using technical vocabulary such as 'directly proportional' or 'linear relationship'.
A6	Try to confirm the relationship given in the briefing sheet by deriving the relationship between the two variables or determining the constant.
A7	Discuss how related physics principles apply to your practical work.
A8	Think about your readings and results to try to qualitatively consider the sources of error.
A9	Justify realistic modifications by explaining how they would reduce error and improve your experiment.
A10	Calculate uncertainties. For example, calculate percentage uncertainty from the spread of repeat readings or use error bars to get uncertainty from two gradients.
A11	Write a conclusion that is based on your data and is related to the aim given in the briefing sheet.
	Maximum marks for this section — 11

The total number of marks available for this unit is 40.

Examiner tip
You are given a blank copy of the marking grid to refer to as you work. It will probably help if you do your work roughly in the order of the criteria.

Summary

● The tables on pp. 10–12 lay out what you need to do to get each mark. However, you have to apply the criteria to the work you are carrying out and its context.

The five sections

The five sections of Unit 3 are:

- the case study or visit
- the report on the case study or visit
- planning the experiment
- implementation and measurements
- analysis

The 2 marks for the report do not refer to the plan or the analysis; they are for how well you write your report on the case study or visit.

We shall look more closely at these sections using some examples to see how the criteria work when applied to practical physics.

Case study

This is a chance for you to research a topic. Your teacher will give you a briefing sheet with the title of your topic and some background to set it in context. You should follow through the example given below.

Exemplar briefing for a Unit 3 case study

What follows explains what you must do for your case study. Remember that you will have a copy of the criteria on a marking grid to help you complete all the requirements. There are 2 marks for the use of clear and correct English.

You must work alone on this assessment.

Background

Geophysics applies knowledge and techniques to the study of the Earth on both the large and the small scale. On the small scale, it is used to investigate archaeological sites and to test the ground before new building work is carried out.

What you are to do

Research two methods of exploring areas of ground either before new building work commences or before archaeological sites are explored.

Discuss the two methods you have found, including the physics principles and explain how these are put into practice.

The practical work you will be carrying out is entitled: *Measuring resistivity to identify a sample of metal*.

> **Knowledge check 1**
>
> What is the formula for resistivity? Explain why the unit of resistivity is the Ω m.

Your work

The first thing to do is look on the internet. But you must remember that for **S1** you will need to find three *different* types of source and any internet site is only one type of source, so you would normally be allowed to quote only one website. The exception is that you are allowed to quote from the website of one of the national institutions, or use data from a manufacturer's website as a second source if this is the only place where they publish such information. You would do well to visit a library where you will find a further source, which might be a physics textbook or a journal published by one of the institutions (such as Chartered Surveying or Civil Engineering); you might also find helpful a magazine such as *New Scientist*. Certainly you will find resistivity as a topic in a textbook and although a great deal of information on the web is of varying quality, you should now be able to find enough to set out the basis for your case study.

When you identify your sources you must give:

- the title, publisher and author of a book
- the title, page number, author, issue number and date for a magazine
- the full web address of a website and the date you visited it

An indication of the source of a website is always helpful — if it is one of the institutions for example.

If you provide full references for all three sources, you should now have the mark for **S2** as well.

You should now do your research. This means visit the websites and read the books and magazine articles you have chosen, making a note of the physics they discuss. Then decide on what you will include in your report. You will probably need to spend about an hour on your research.

Examiner tip

Make sure that your technical vocabulary is correct throughout your case study so that you do not lose **S5**.

Now, write your report. Start with the topic title and amplify what is said in the briefing sheet to describe the topic in some detail. It is a good idea to include something about the relevant physics. You must make sure that you are using the technical terminology correctly.

In this case study here you would describe the factors that determine how the resistivity of the ground varies; it is by measuring the resistivity that the physicists are able to construct a picture of what lies under the ground without having to dig it up and perhaps spoil the archaeological site before they have even seen it. You then describe the physical principles of measuring the resistivity and then you are well on your way to **S3**, **S4** and **S5**. This section should be quite detailed as it sets the foundation for the next one.

For **S6**, **S7** and **S8** you will write about the benefits of using the physics and how this technique makes the process so successful; you should quote a specific example to strengthen your point. You should include some hard data about your study, and since the briefing sheet asked you to research *two* methods you might find you need to describe two different sources. It is more important to discuss the physics than becoming bogged down in detailed applications.

You can make a point of drawing the physics clearly to the focus of your case study so that you complete this section neatly. Another example might be the physics of solar cells or indeed almost anything *you* would like to research — ask your teacher. The good thing about the case study is that it enables you to look at a wide variety of applications.

For **S9** you have to link the case study to the practical. In this example, the briefing sheet does not mention resistivity, only geophysics. Therefore, your link is simply to say that resistivity has many applications, for example identifying a sample of a metal.

Remember you are allowed to write only 600 words. However, it is a good idea to include some diagrams and if you are using ICT in writing your report then you could include some material from the websites you visited — properly referenced of course.

There are two marks, **R1** and **R2**, for the written quality of your report. Make sure you spell all the technical terms correctly and write clear sentences that can be understood easily. You should structure your report by including subheadings.

Visit

Your teacher might be able to organise a visit for you but this depends on your local circumstances. If you do go on a visit, look carefully for all the physics there — some of it might not be so easy to see. Take a notebook with you, as to gain **S6** you must come back with some data. There might be some information available as a booklet or similar that you could also bring back with you. Going on a visit covers **S1**.

Now write your report. Start with a statement of where you went and give some details of the reason for your visit. A suitable visit could be to a small industrial firm nearby who research and make a product with some technical aspect, for example car parts or machinery. It could be to a particular hospital department; you should state which department and give the reason for going. Then go on to describe the details of the visit. It is here that the technical aspects should feature, so you might describe what you saw at the industrial location or say that you went to see the X-ray machine to find out about doses and damage to tissue. This covers **S2** and **S3**, which are about the specific details of where you went.

For **S4** and **S5** you have to use technical terms to describe the physics principles behind what you saw. This is likely to be a long way off the specification, but the basic principles should be clear and you should include a brief description in your report. So if you visited an X-ray department an outline of how the X-rays are produced and what governs their ability to penetrate materials would be appropriate. If the industrial firm produces plastic lens systems for car lights then some mention of refraction and colour transmission is appropriate.

For **S6, S7** and **S8** you should describe what you saw and heard. There is always a guide, so the context of the physics is explained together with how the physics principles help the people involved to achieve a better result in whatever they are doing. You will also get some data from your guide. For the hospital visit you might discuss the effect that X-ray technology has had on the ease with which doctors can diagnose problems or you might discuss the health implications for the people who work there. In the industrial case you could consider the cost advantage and the

Examiner tip

Make a list of three topics that you might like to do as a case study — they could be either physics topics or applications.

Examiner tip

Keep your report to 600 words and remember to use subheadings.

ability of manufacturers to ship and fit spare parts. It all depends on your particular visit, so enjoy it!

Remember you are allowed to write only 600 words. However, it is a good idea to include some diagrams and if you are using ICT in writing your report then you could include some material from the websites you visited — properly referenced of course.

There are two marks, **R1** and **R2**, for the written quality of your report. Make sure you spell all the technical terms correctly and write clear sentences that can be understood easily. You should structure your report by including subheadings.

Planning

You are allowed to make the link between your case study or visit and the practical work either at the end of the first section or before your plan.

The ideas behind the planning criteria are discussed below together with what you must do to satisfy these. Here the criteria are linked to two particular practicals, but what is said applies in general to all the criteria. The practicals are the resistivity of a metal in the form of a wire and the Young modulus of rubber. These are by no means the only practicals at this level but they are sufficiently different to demonstrate how the criteria apply widely. In practice, students submit a huge variety of practical work.

The most important thing to establish is the aim of the experiment. You must be clear what the aim is and understand what you are trying to measure.

The practicals

The aim is the most important thing to establish. For the two practicals chosen the aims are as follows:
- **Experiment A**: To identify a metal by measuring the resistivity of a wire made of that metal.
- **Experiment B**: To measure the Young modulus of the rubber in a rubber band.

So, to achieve the aim, it is necessary to take measurements in order to calculate a value of a specific constant, which can then be compared with the accepted value.

Apparatus

List and diagram

P1 requires you to produce a list of all the apparatus you will need to take all the necessary measurements. Your list must include both the item under investigation (here a piece of wire or a rubber band) and the means to experiment on it (here a power pack or masses). For both experiments you need a micrometer to measure the width of the item. Remember to include everything. A diagram will help, so this is a good place to draw your diagram, even though this is for **P13**. Make sure that your diagram is clear and shows what you will actually do. For example, you should indicate lengths on your diagram with dotted lines across to a rule positioned close by.

Your diagram should show the apparatus assembled as a scientific diagram *not* an illustration. This means, for example, that there is no need to draw a multimeter — it is enough to show the correct circuit symbol in the correct place in the circuit. Similarly, you are not expected to draw a micrometer screw gauge; it is enough to say that you will use one.

Apparatus list for Experiment B

- Rubber band — about 10 cm long
- Retort stand, two bosses and clamps
- G-clamp
- 7 × 100 g masses
- 1 × 50 g mass
- Mass hanger
- Metre rule
- Micrometer screw gauge

Examiner tip
Notice that some information is given about the rubber band and about the masses. This is to help the technician provide the apparatus you want.

The diagram for Experiment B is quite straightforward (Figure 1). For Experiment A the diagram must show a simple electric circuit diagram and the physical arrangement of a metre rule and attachments.

Knowledge check 2
Write a list of the apparatus needed for Experiment A: to identify a metal by measuring the resistivity of a wire made of that metal.

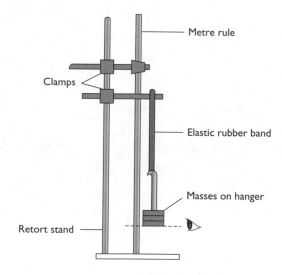

Knowledge check 3
Draw a diagram for Experiment A to show how you would set up the apparatus listed in answer to Knowledge check 2.

Figure 1 Apparatus for measuring the Young modulus of the rubber in a rubber band

Instruments

For **P2** to **P5** you have to select the measuring instruments you will use and say why they are suitable. Remember that they do not need to be for the variables.

In Experiment B you would use a metre rule to measure the length of the rubber band because it has a precision of 1 mm and will be long enough to measure the fully stretched band, probably 30 cm. This will enable you to measure the original length of the band, the extended length and subtract to find the extension — this is called a difference method. The metre rule is an easy choice since no other piece of common laboratory apparatus has a scale long enough. If you accept that the masses are labelled correctly then the other measurement to be made is the cross-sectional area.

Knowledge check 4

Explain why a micrometer screw gauge is the best instrument to measure the diameter of a wire.

A micrometer screw gauge is a better choice than callipers because the micrometer is more precise. It can measure to 0.01 mm or even 0.001 mm if it has a digital readout. To justify your choice you should discuss the precision and range of the instrument with reference to the measurement to be made, which you may have to estimate.

In Experiment A you will have to measure the resistance of the wire — a few ohms — and the length and diameter of the wire.

Techniques

For **P6** you must show that you plan to use the instruments well. Some examples are given in the table.

Measurement	Instrument	Technique(s)
Length of rubber band	Metre rule	Place rule close to band Lay the rubber band out straight without stretching it Read the position of each end of the band and subtract
Diameter of a wire	Micrometer screw gauge	Take a pair of readings at right angles to make sure it is circular and find the mean Take readings at several places down the length and find the mean
Time for a model car to roll down a ramp	Stopwatch	Draw clearly defined start and stop points on a ramp Make sure the car starts from rest Repeat and find the mean
Temperature of water in a water bath	Thermometer	Place thermometer close to object being heated Stop heating before reading Stir water
Potential difference	Multimeter	Use the most sensitive voltage range Check connections are secure
Length of a simple pendulum	Metre rule	Clamp rule alongside Record the position of the bottom of support (top of string) and the centre of mass of the bob Read the rule with your eyes at the same horizontal level as the position recorded
Diameter of a test tube	Callipers	Repeat readings in different orientations Repeat readings at different places down the tube

Knowledge check 5

What would you use to measure the time period of a simple pendulum? What steps would you take to make your measurement as accurate as possible?

Examiner tip

Always say exactly what you will measure. To find a cross-sectional area you have to measure the diameter — you cannot measure the radius.

Knowledge check 6

What would you use to measure the electrical resistance of a circuit component? What steps would you take to make your measurement as accurate as possible?

This also covers **P9** since you must comment on whether you will repeat your readings. The comment should include the reason(s) why repeating readings will improve them — it enables anomalies to be spotted and removed before calculating a mean, which itself reduces random error. You cannot repeat readings where time is one of the variables since you have to stop the increase to repeat a measurement. When temperature is one of the variables you should do just that, i.e. remove the heat source for a short while to allow thermal equilibrium to be reached — everything must be at the same temperature.

Your teacher will read your plan to make sure that your experiment is safe. For **P10**, you should say why it is safe. If by any chance this is not the case, then you must identify any hazard(s) and describe suitable precaution(s). In Experiment A the emf produced by the ohmmeter is not high enough to cause a shock and in B the wire is not under tension, so no precautions are necessary. If the experiment is safe you should identify any possible hazards and explain why you think no precaution is necessary.

The uncertainty in your result — remembering the aim of the work — might be caused by the way you use the apparatus. For **P12** you have to identify the source of the uncertainty. For example, in Experiment B there was no check that the masses were exactly 100 g and the rubber is easily squashed by the micrometer as you tighten the screw. Taking the uncertainty in one reading on a metre rule to be ±1 mm (the precision), then the extension has an uncertainty of ±2 mm since you are reading at both ends of the rubber band. This might be a significant percentage of the extension.

Method

Write out your method using bullet points. Make it a clear and logical progression — imagine you are writing instructions for one of your friends to follow — and you will earn **P14**. You should identify the dependent and independent variables and say how you will use your data to find the answer to the question in the briefing. You should describe the graph you will plot and how you will use this to find the value required by the briefing sheet. This gets you **P7** and **P11**.

Finally, for **P8** identify any other factors that might affect the outcome and state how you will control these variables. You should not be too ambitious here but consider only those things that are likely to have a real effect on your work. Room temperature, for example, is unlikely to have an effect; there might be a difference between summer and winter but it will not change much during the course of your work and you need not mention it. There are probably no variables to control in Experiment B. Nothing will affect the length (dependent variable) of the band apart from hanging masses on the end (independent variable). In Experiment A the wire might get hot if you are using the two-meter method that needs a power pack, but a simple ohmmeter produces a current that is too small to have this effect so, again, there are no variables to control — unless pushing hard to make good contact with the wire changes the shape of the wire, but this is not a variable — more a point about your technique.

You have now written your plan and it is time to get on with the practical work.

Implementation and measurements

You should follow your plan but keep in mind what you are trying to find out — the value of the quantity asked for on the briefing sheet. First, for **M1**, draw up a table and head each column with the quantity and the unit, for **M2**. For length in metres the column heading is l/m — but you must be careful to record the length in *metres*. It might be easier to use a sub-unit if that is what the instrument reads, so you might record length in centimetres (l/cm).

Knowledge check 7

Comment on the safety aspects in Experiment B: to measure the Young modulus of the rubber in a rubber band.

Examiner tip

Think about why your experiment is safe — you must write down your reasons for **P10**.

Knowledge check 8

Identify the sources of uncertainty in Experiment A: To identify a metal by measuring the resistivity of a wire made of that metal.

Examiner tip

The independent variable is the factor that you vary (or time, if that is one of your measured variables). The dependent variable is what you measure after having changed the independent variable.

The independent variable is nearly always plotted on the x-axis and the dependent variable on the y-axis.

Knowledge check 9

Identify the independent variable and the dependent variable in both Experiment A (resistivity) and Experiment B (Young modulus).

Examiner tip

Read through your plan carefully and imagine one of your friends is going to do the experiment using your plan. Does it make enough sense or would they need to ask you further questions? Like 'what rubber band?'.

Significant figures

You are going to record what your instruments read and this is usually to 3 significant figures (s.f.), but could be 2 or 4. If you are reading potential difference (p.d.) on the 2 V range of a multimeter and it varies from below 1 V to above 1 V then the readings you record might be from 0.736 V to 1.598 V. Record these to 3 s.f. and 4 s.f. respectively. You should use the least number of significant figures when calculating values for derived quantities, so this p.d. should be used to 3 s.f. in calculations. Your graph should be as precise as possible. In general, 3 s.f. are needed for plotting the points, so gradient calculations should also be given to 3 s.f. You can go wrong by not using the **precision** discussed in your plan. If you said the rule measures to a precision of 1 mm then you should record readings to 1 mm, not 1 cm.

The correct number of significant figures is the number shown on your measuring instruments.

Readings

For **M3**, you need to take enough readings to plot a reliable graph. If you expect your graph to be a straight line then you should have six readings. For a curved graph you should take more readings where the graph curves and the readings change rapidly, and take fewer readings where it is straight.

For Experiment A, a straight line is expected from the theory, so if you take readings every 10 cm from 50 cm to 100 cm you will have six readings over a reasonable spread, with the length doubling from 50 cm to 100 cm. You do not want the wire to be too short as the shorter the wire the smaller the resistance and the greater the current, which might just heat up the wire. Keep an eye on the readings to check that they are going the way you expect.

For Experiment B, a straight line is not expected — rather a graph with three straight(ish) sections. You do not want to take dozens of readings so take more where the graph curves and fewer where it is straighter — this will get you **M4**. You need to keep an eagle eye on those readings!

The point is that you do not always need to use equal increments of the independent variable.

Analysis

First, plot your readings on a graph. You must do this by hand without using a software programme. You then evaluate your findings and move towards a conclusion in response to the briefing sheet.

Graph

A1 is for having the correct axes, as described in your plan, and for giving the axes the correct units. The gradient of a graph has no units because the axes are labelled with units and, therefore, you plot pure numbers — hence no gradient units. So if you are plotting a length, l, measured in millimetres you should label the axis l/mm; if plotting the time period squared, measured in seconds squared, then the axis label is t^2/s^2. You should have a good reason for plotting the independent variable on the y-axis because it usually goes on the x-axis.

A2 One purpose of a graph is to display data. The data points should occupy at least half of both the axes of your graph; you should not include the origin unless you can do so without confining the points to one corner. The other purpose of a graph is to enable you to take readings from intermediate points (interpolate), so make sure your scale is sensible and easy to read — usually 1, 2, 4 or 5 units per cm of graph.

A3 Plotting should be accurate to the nearest millimetre.

A4 The best-fit line should have points above it and points below it and should not necessarily pass through the origin. If you are plotting a current–voltage graph then you should think of the origin as another data point. The graph might not be straight over the whole data range, in which case draw a straight line as far as it fits and subsequently draw a curve. It may be that your mathematical model does not fit the data and the points really do follow a curve. If the top and bottom points are below the line and the middle points are above the line then the graph should be drawn as a curve. You should draw what you see and not what you were expecting — there might be a reason why your data do not follow the model.

You should describe your graph using appropriate vocabulary — not 'correlation', which suggests a rough link between the variables and means that your data are probably not very good. If the graph is a straight line through the origin then the two variables are directly proportional; if the y-intercept is not zero then you can say there is a linear relationship between the variables. You should also note how close the points are to the straight line you drew. Then calculate the gradient of this line to determine the constant. You now have **A5** and **A6**.

Figure 2 on p. 22 is a graph of resistance against length for a metal wire. Although the uncertainty in the length might be ±4 mm, the data have been plotted without error bars on either axis. The line of best fit is drawn passing through the crosses and is labelled BB — notice that it has been extended to the edges of the grid.

Error and uncertainty

Error in a reading may be systematic or random. The value of a derived quantity will have an uncertainty that comes from the readings used in the calculation; the uncertainty reflects your confidence in your final result.

Error

A **systematic error** is caused when something affects all of the readings of an instrument in the same way. In Experiment A the meter might not read zero when the connecting leads are joined together, probably because the leads have some resistance. If this is not negligible then each meter reading will be too high. A zero error might occur in Experiment B if the position of the top of the rubber band is recorded incorrectly.

A **random error** results from variability in what you are reading. For example in Experiment A, the wire might not be completely circular throughout its length — it might be worn — and so the diameter reading will vary, showing random error.

Examiner tip

You do not have to include 0,0 (the origin) on your graph.

Examiner tip

When candidates choose a difficult scale they usually make a mistake in plotting the points and lose **A3** as well as **A2**.

Knowledge check 10

Take measurements from the graph in Figure 2 and calculate the gradient of the best-fit line, labelled BB.

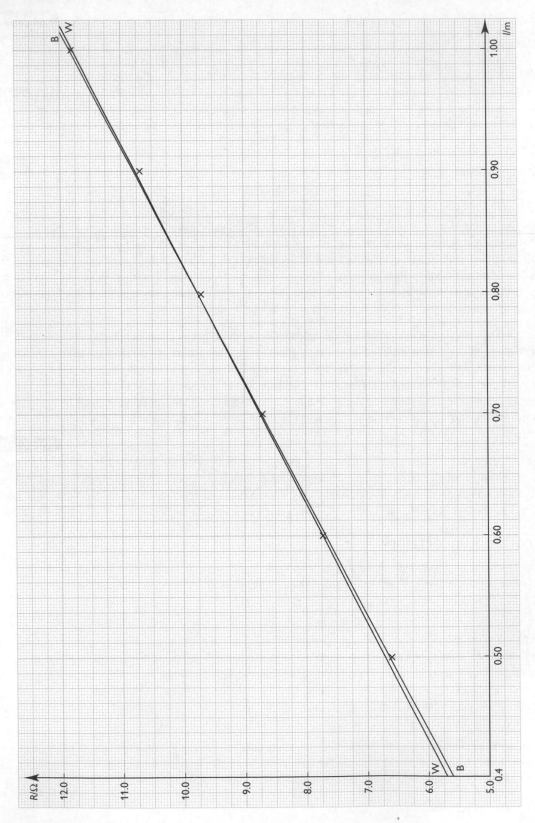

Figure 2 Resistance against length for a metal wire

Random error is shown by the spread in repeated readings, which might also come about because what you are reading is difficult to measure — think about measuring the height to which a table-tennis ball rises after being dropped onto a hard surface.

For **A8** you have to think about where error might have arisen in your experiment. In Experiment A there should be little error in the readings because you can take your time over them and recording the position of a pointer on a metre rule is as precise as the scale — ±1 mm. You are quite likely to get a systematic error from the resistance of the leads, but this can be measured and taken into account. In Experiment B you might find that the rubber band creeps — continues to extend slowly under constant load — making it difficult to decide on the reading. The micrometer might squash the rubber giving a false value for the area. You should be realistic in considering these errors; if you consider them to be very small then say that there is little error and explain why you think this.

Uncertainty

Uncertainty arises when you consider the effect of the errors in the context of the actual measurement. There are two ways to take this into account.

- If there is a spread of readings then you can take the range of the spread and divide by the mean value to find the percentage uncertainty in a value. This is likely to be a modest uncertainty but not insignificant. If it were the same in both variables the overall uncertainty might be 10%, which is verging on significant.

- The other way to consider uncertainty is to use the graph. Plot error bars on the graph — these are lines drawn vertically and horizontally from the plot and indicate how much error there is in the reading. Now draw a best-fit line through the data points on the graph and then draw a worst-fit line that just fits through all the error bars. The uncertainty in the gradient is given by the difference between the two gradients. It can be expressed as a percentage of the gradient of the line of best fit.

Conclusion

The aim of the uncertainty calculation for **A10** is to give a measure of the confidence you can have in your conclusion, **A11**. The conclusion is the answer to the question in the briefing and the uncertainty, informed by your thoughts on errors, will give your results some validity. At some stage you will have mentioned the physics behind your practical. For example, you might describe the theory or use some part of it to explain why your observations follow the pattern that they do. This contribution, for **A7**, can be included anywhere in the write-up, from plan to analysis.

As part of your overall conclusion you need to think about how you might improve the experiment, for **A9**. Any modifications you suggest should be practical and work in your laboratory without requiring exotic equipment, although you could suggest equipment your school does not have. Your suggestions should reduce uncertainty and improve your final answer in terms of accuracy or precision. It is often possible to do this with a data logger but you must say how you would use it and why it improves readings. To obtain this mark you will have to write at least a couple of sentences.

Knowledge check 11

A student measures a variable and finds the following 6 values:

4.76, 4.88, 4.65, 4.75, 4.72, 4.74

The mean value is 4.75 but what is the effect of the random error? Calculate the percentage error in this measurement.

Knowledge check 12

Calculate the gradient for the worst-fit line, labelled WW, on Figure 2.

Examiner tip

Finish your write-up with a clear statement of what you have found, related clearly to the briefing; this is the conclusion.

Worked examples

Planning is the skill that needs the most building up. Here are some examples of experiments that you should meet during lessons. The questions are designed to get you to practise the planning you need to do in order to score marks.

g by free fall

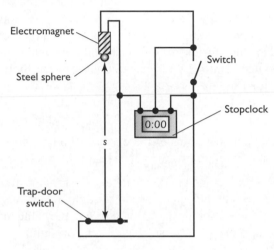

Figure 3 Simple apparatus for measuring *g*

This experiment (see Figure 3) measures the time taken, *t*, for the ball to fall through a distance, *s*, in order to obtain a value for the acceleration due to gravity, *g*. Since there is not a linear relationship between the variables this practical is not suitable for a full assessment at AS, but there are some useful questions that can be asked.

Q What instrument would you use to measure the distance fallen, *s*? P2

Examiner tip
It is always a good idea only to answer the question that has been asked — you do not need to make your work harder.

A A metre rule (not ruler).

Q How do repeat readings help you get a better value for *g*? P9

A You should take at least three readings for *t* for each value of *s*. Repeating readings helps to spot anomalies that will not be used. They also make it possible to calculate a mean, which will help to take account of random error. The values found for *t* will be more accurate.

Q When you measure *s* how will you avoid introducing a systematic error? P12

A Measure to the bottom of the sphere — as shown on the diagram — since the clock stops when the bottom of the sphere hits the trap door.

Q What will be the main source of uncertainty? P12

A Probably the time t, since it is such a small interval. $\Delta t/t$ will therefore be quite large.

ⓔ The mathematical model is $s = ut + \frac{1}{2}at^2$. So we must make sure the sphere falls from rest — u is zero. A graph of s against t^2 will be the easiest — the independent variable on the y-axis for a change. How will you use the graph to find a value for g? There are no variables that need controlling so for **P8** you need to say that this is the case.

Viscosity

In an experiment to determine the viscosity of some oil at room temperature, a steel ball-bearing is released from rest just below the surface of the oil and timed as it crosses a series of equally spaced marks on a glass cylinder This is illustrated in Figure 4. The experiment is repeated three times, and the average time, Δt, taken for the ball to cross each 5 cm division is calculated. The ball-bearing will reach terminal velocity, enabling the viscosity of the oil to be calculated.

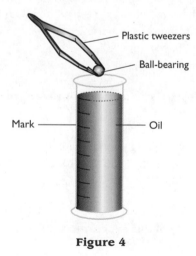

Plastic tweezers

Ball-bearing

Mark

Oil

Figure 4

Q The diameter of the ball-bearing is about 3 mm. What instrument would you use to measure it? Justify your choice. P2 and P3

A A micrometer screw gauge because it has a precision of 0.01 mm, which is small compared to 3 mm.

Q What would you use to measure the time taken by the ball-bearing to cross each 5 cm division? Justify your choice of instrument. P2 & P3

A A stopwatch with a precision of 0.01 s, which is small compared to the likely measurement. Use the lap timer facility to 'freeze' the display — this makes the reading easier to record.

Q Discuss how you would make your measurements of the time as accurate as possible. P6

A Time a fall from the top to each mark separately, repeat this and find a mean for the time to reach each mark. (This assumes you can retrieve the ball-bearing each time, usually using a strong magnet).

Q Discuss the variables you might control to ensure that this is a fair test. P8

A It is only fair if the ball-bearing falls from rest each time, so you must control this in your method. Despite how difficult it might be to retrieve the ball-bearing you should always use the same one. Viscosity varies with temperature; so although you cannot control room temperature you might measure it before and after your work to see if it has changed.

Q Comment on how repeat readings will improve accuracy. P9

A Repeat readings help in spotting anomalous readings, which will not be used. Taking a mean of the other readings will help to take account of random error. The values for t will be more accurate.

Q Comment on the safety of this experiment. P10

A There are no hazards in this experiment so no precautions are necessary. You should take care not to spill any oil and mop up as quickly as possible any spills that do occur.

ⓔ This experiment can get a bit messy as every time you retrieve the ball-bearing you get some oil with it. Draw a sketch graph of distance against time for the ball-bearing and try to explain how you will know from your measurements that the ball-bearing has reached terminal velocity. How would you calculate that terminal velocity?

How could you do this experiment at a different temperature?

Internal resistance of a cell

In an experiment to find the internal resistance of a cell the circuit shown in Figure 5 was used.

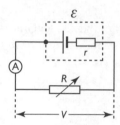

Figure 5

The resistor R was varied and readings were taken for the potential difference, V, and the current, I.

Q What instrument would you use to measure the potential difference, V? Justify your choice of instrument. P2 and P3

A   A multimeter switched to a volts range. By selecting the smallest range a high degree of precision can be achieved. For a few volts, the precision will be 0.01 V and this is small compared to the value measured.

Q Discuss the variables you might control to ensure that this is a fair test. P8

A   There are no other variables. You might ensure that the resistance does not become too small. A large current will heat the resistor and could also affect the cell.

Q Comment on whether repeat readings will improve accuracy. P9

A   There is no point in repeating readings because nothing changes between readings. Looking away and looking back at the meters makes no difference. You could take more readings than the usual 6, thus improving reliability if your graph is a straight line.

Q Comment on the safety of this experiment. P10

A   Since the voltage and current are low, there are no hazards in this experiment and so no precautions are necessary. You should ensure that the current does not become too high so that no component becomes hot.

Q How will the data be used? P11

A $E = V + Ir$ where E is the emf of the cell and r is its internal resistance. Plot a graph of V against I and the gradient will give a value for $-r$.

ⓔ So $V = -Ir + E$ and V is plotted on the y-axis and I on the x-axis. The slope is $-r$ (note that it is negative) and the aim of the experiment is to find the value of r. If r is constant over the range of readings then the line will be straight.

Q Identify any sources of uncertainty. P12

A   There are no sources of uncertainty other than random variation and that caused by the precision of the meters.

ⓔ This is an example of an experiment where repeat readings are not helpful. Other examples involve taking readings at certain time intervals or at specific temperatures.

ⓔ **In all these experiments do not invent hazards or uncertainties just because you are asked about them. You should say why there are none, but you must say something because a blank page cannot score any marks.**

Unit 3 exemplar

Visit report

A student was taken on a visit to Thorpe Park, a theme park in Surrey. The following report was submitted.

Account of visit

On Monday the 23rd we went on an excursion to learn about the physics involved in the thrills of rollercoasters. Along with enjoying what was on offer we were given a lecture on how physics is used and applied in the making and daily use of rides.

The day started with a tedious bus to Chertsey, Surrey. After being crammed into a minibus we couldn't wait to get off. Once at the theme park, being a person that doesn't like rollercoasters, I immediately thought of what rides I could go on. It wasn't long before I was thrown in at the deep end and already on the rides, where one came after another. The day was wet and cold, so I'm not sure water rides was a smart idea, but it was a day to have fun!

After a brief period to warm up, we went to a lecture where I learnt about the rides at Thorpe Park. I found out that on 'Tidal wave' the boats are monitoring acceleration due to gravity for health and safety and that the viscosity of water is enough to keep the boats moving.

All the machinery at Thorpe Park is run by nice clean electricity — there are no smelly diesels — and all the energy is supplied through underground wires. The place is an attraction for thrills due to gravity (G force) — the twisting and turning, ups and downs, all contribute to that stomach-in-mouth feeling. Physics enhances the feel by a force of 4.2 Gs acting on you, gives you a thrill you can't get at other places.

Thorpe Park uses huge amounts of electrical energy and my practical will be about resistivity as that affects how hot the wires get and how much energy is wasted.

Ref		Commentary on the visit report
S1	✓	Attends visit to look at the physics of rollercoasters
S2	✓	Describes venue as a theme park with a variety of rides
S3	✓	Describes lecture and topics covered
S4		No relevant physics principles are discussed, viscosity is mentioned but the application is not explained. The student should have said how viscosity works to apply a force to move the boats.
S5		Technical terms are not used. 'Twisting and turning' should be related to circular motion and acceleration.
S6		The relevant piece of information is 4.2 Gs but this means nothing by itself. It would be better to say 'the acceleration required a force of 4.2 times normal gravity' or something similar.

Ref		Commentary on the visit report
S7	✓	The energy is supplied by electricity, which is cleaner than having diesel engines alongside the rides and the cables are hidden.
S8		There is not enough about the environmental improvement to score this mark. There is a cost benefit in using centrally generated electricity (so long as the cables don't get too hot) and any pollution can be dealt with more effectively, but this is not mentioned.
S9	✓	There is a clear link to the practical work on resistivity.
	5	Marks scored
Ref		Commentary on the report marks
R1	✓	The report is written well enough and is easy to follow. There are few technical terms but these are spelt correctly.
R2		There are no subheadings and little evidence of a structure other than that implied by the order of the criteria.
	1	Mark scored

ℯ The student obviously enjoyed the trip but at about 300 words there is not enough opportunity to write enough to access all the marks. There are few hard data in this report so it might be that the student did not have any paper to make notes during the lecture. Remember — it is a physics visit, even if it is fun!

The student then went on to the practical work and produced the following report.

Practical work

Investigating the resistivity of constantan

The student was told the aim of the experiment is to find a value for the resistivity of a piece of wire to see if it is constantan, which has an accepted resistivity value of $5.00 \times 10^{-7}\,\Omega\,m$.

Planning

Apparatus

- A metre ruler (accurate to ±1 mm)
- A micrometer
- Two cells (total 3 V)
- Wire
- Crocodile clips
- Connecting leads
- Ammeter (measure current accurate to 0.01 A)
- Voltmeter

Method

(1) Set up the experiment as in Figure 6.

(2) Place the metre rule alongside the wire.

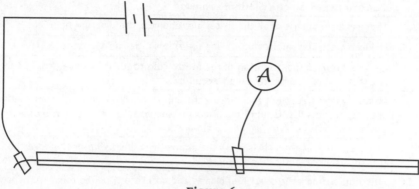

Figure 6

(3) Using the rule to measure distance, place the crocodile clips 15 cm apart on the wire (this will be to 1 mm accuracy).

(4) Read off the values on the ammeter and voltmeter and record them in a table.

(5) Switch off the circuit between each reading to ensure the wire does not heat up, as this could become a burn hazard.

(6) Repeat this reading twice more and find an average value for voltage and current.

(7) Once an average reading has been taken, repeat steps 4, 5 and 6 at 15 cm intervals between the crocodile clips.

The length of the wire is the independent variable. Both voltage and current are dependent variables and to make the experiment a fair test only two cells are connected in the circuit.

- Draw a graph of resistance (Ω) on the y-axis and length of wire (cm) on the x-axis (Figure 7).
- Using a micrometer, measure the cross-sectional area. Grasp the wire slowly so as not to cut into it. Do this at three points on the wire and take an average for the whole wire because it might vary.
- From the graph, calculate the resistivity by using the formula 1/gradient = resistivity, then compare the answer with the value given.

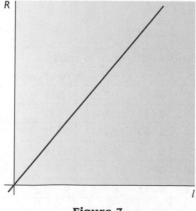

Figure 7

Ref		Commentary on the planning marks
P1	✓	There is a simple list of all the materials required, the diagram helps.
P2	✓	Makes a clear statement about using the rule to measure distance.
P3	✓	Quotes 1 mm as the precision of the metre rule to justify choice. Ignore the fact that this is incorrectly called 'accuracy'.
P4	✓	Makes a clear statement about using the ammeter to measure current — this is in the apparatus list. Could have discussed the voltmeter or the micrometer for this mark.
P5	✓	Quotes 0.01 A as the precision of the ammeter to justify choice. Ignore the fact that this is incorrectly called 'accuracy'.
P6	✓	Use of micrometer is described clearly and accurately, taking measurements at different places along the wire and not cutting the wire with the jaws. Ignore that the student plans to use it to measure area.
P7	✓	Makes a simple statement that the length is the independent variable and the voltage and current are both dependent variables.
P8		The two cells do not make this a fair test. Keeping the temperature of the wire constant would be better because temperature affects the resistivity directly, but this is not mentioned.
P9	✓	Repeat readings are suggested for all three measurements and a mean will be taken. The variation in diameter is mentioned as a reason.
P10	✓	Safety is considered in the event of the wire becoming very hot. It would be better if it had been mentioned that this would happen if a short length of wire were used. Mark awarded on the basis of 'benefit of the doubt' (BoD).
P11		The appropriate graph was selected but used wrongly. The gradient is not the inverse of the resistivity. The cross-sectional area has to be included.
P12		There is no mention of uncertainty or error.
P13		The diagram is not appropriate for this level. The crocodile clip is not at the zero mark on the rule, but there is no voltmeter so the circuit cannot be used to measure the resistance.
P14	✓	The plan is clear and methodical.
	10	Marks scored

ⓔ The student went about the work in an organised fashion, for the most part following the order of the criteria. The more demanding marks were not scored.

Results

Length in mm	Current in amperes				Potential difference in volts				Resistance in ohms
	A_1	A_2	A_3	Average A	V_1	V_2	V_3	Average V	
100	1.05	1.05	1.06	1.05	0.518	0.519	0.520	0.519	0.49
200	0.66	0.67	0.67	0.67	0.630	0.630	0.631	0.630	0.94
300	0.47	0.47	0.47	0.47	0.682	0.683	0.681	0.682	1.45
400	0.36	0.36	0.36	0.36	0.715	0.714	0.713	0.714	1.98
500	0.31	0.31	0.32	0.31	0.728	0.726	0.725	0.726	2.34
600	0.27	0.27	0.27	0.27	0.744	0.743	0.744	0.744	2.75

Diameter of the wire in mm = 0.37, 0.37, 0.38   Mean diameter = 0.373 mm

Ref		Commentary on the implementation and measurements marks
M1	✓	Good table that includes all repeats. The values given for the current are to the precision quoted in the plan. Ignore the calculated values, resistance and area.
M2	✓	The units in the table are correct, even though they are different from those in the plan. The student has changed his/her mind.
M3	✓	Six measurements are fine since the graph is expected to be straight.
M4	✓	It is odd that the range does not go up to 1.0 m since the rule is 1 m in length; but the resistance changes by a factor of 5. BoD.
	4	Marks scored

ⓔ The student need not repeat all the readings since nothing changes between taking them. The spread of readings is small, showing a small random error, as you might expect.

Analysis

From my results (see Figure 8 on p. 34) I can see that the resistance increases when a longer length of wire is used. The longer the length of the wire the higher is the resistance of the wire. This is due to fewer charge carriers over the distance.

To calculate the resistivity:

$$\text{gradient of graph} = \frac{\Delta y}{\Delta x} = \frac{2.79 - 0.51}{0.6 - 0.1} = \frac{2.28}{0.5}$$

$$\text{gradient} = 4.56\,\Omega\,\text{m}^{-1}$$

resistivity $\rho = \frac{RA}{l}$ where $\frac{R}{l}$ is the gradient of the graph

so, $\rho = \text{gradient} \times \text{cross-sectional area}$

$$\text{cross-sectional area} = \pi r^2 = \pi \times (0.373 \times 10^{-3})^2 = 4.37 \times 10^{-7}\,\text{m}^2$$

so, $\rho = 4.56 \times 4.37 \times 10^{-7} = 1.99 \times 10^{-6}\,\Omega\,\text{m}$

The graph passes close to the origin, which shows that there was very little resistance when the length was zero. It should go through the origin, so there must have been a small resistance in the connecting leads. The wires did not get hot because I turned off the current between readings. Because I repeated my readings and all the points are close to the line, I think the results are good.

To improve my results I would replace the ammeter and voltmeter with an ohmmeter as it would automatically measure resistance and reduce my percentage chance of errors. I would also apply extra tension to the wire to remove kinks in it. This should give me a better reading for the length of my wire because it will be straighter than before.

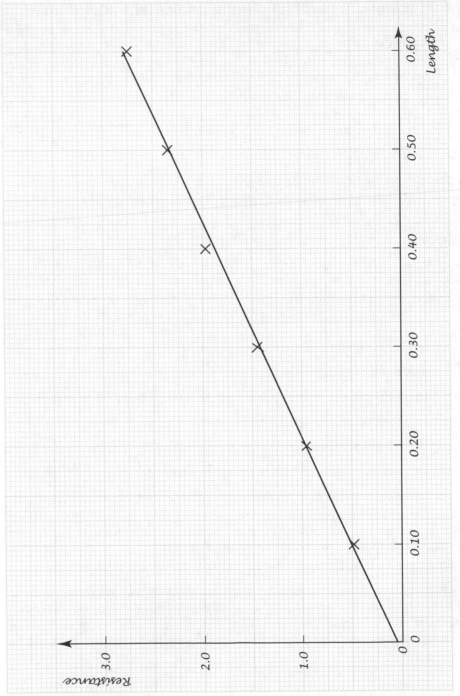

Figure 8

Ref		Commentary on the analysis marks
A1		The axes are labelled but do not show units.
A2	✓	The data points fill the page in both directions.
A3	✓	The points checked are accurate to ±1 mm and there is no penalty for changing the units from those in the plan.
A4	✓	The best-fit line has points above it and below.
A5		The comment that 'the resistance *increases* with length' is not enough at this level. Since the graph passes close to the origin it is reasonable to say that the variables are directly proportional. If some distance from the origin but still a straight line then the comment is that there is a linear relationship between the variables and they increase together.
A6		This student gets very close to this mark, but unfortunately uses the diameter rather than the radius to calculate the area and hence the resistivity, and so loses the mark. A shame, because their value is exactly four times too big, so the correct area would give exactly the figure given in the briefing.
A7	✓	The student does recover from the mistake in the planning and shows that they understand how to obtain the resistivity value from the readings taken. The units are correct. Ignore the comment about charge carriers and award BoD.
A8	✓	There is a comment about the zero error shown by the graph, which is linked to the resistance of the wires, and their temperature is mentioned.
A9	✓	The modifications are both realistic and justified.
A10		There is no attempt to calculate uncertainties.
A11		The conclusion mentions how close the points are to the best-fit line but fails to fulfil the requirements of the briefing by identifying the material as constantan or not, probably because the resistivity value obtained is four times too big.
	6	Marks scored

ⓔ The student scores a total of **26 marks**. In the Analysis section not enough attention was given to the more difficult skills.

A simple calculation error not only lost marks but seems to have put off the student from commenting on the result in **A11**. You must answer the question put in the briefing — here, identifying the material. In this case, the student should say that the material is not likely to be constantan since the resistivity is four times too big. It is always worth checking your calculations by doing them again from scratch. The analysis is too brief, which indicates a lack of thought and this is reflected in the mark.

The student lost 2 marks by ignoring the uncertainties in planning and analysis. The repeat readings for current and potential difference are not helpful as no change is made in repeating them, so the student is simply looking at the meters again — this is not repeating readings. This work would probably get the student a grade C.

Unit 6 Experimental physics

Overview

This unit is a development of the work that you did for Unit 3. It is important to be able to recognise where the demand is greater and what you must do to achieve the higher standard. As for AS there are 40 assessment criteria, each of which stands alone, i.e. you do not have to satisfy one of the criteria before you move on another. As with AS, you will have a clean copy of the marking grid with you as you carry out the work. The work is done during normal class time and although there are no time constraints you will probably need about three 1-hour sessions.

At A2 there is no visit or case study, everything is completed in your physics laboratory or classroom and you must hand in your work each time you leave. The 11 marks allocated to the case study or visit in Unit 3 are split between the three phases of this unit — planning, implementation and measurements, and analysis.

Your aim is to produce a conclusion about your work that is based on your results. The **briefing sheet** gives overall direction and it is important to read this carefully before you start.

Planning

The planning section is worth 16 marks.

The work starts with a **briefing sheet** which provides information about what you will be investigating. It sets the work in context, gives you a framework to follow and provides you with an aim. It is quite a short document that contains a mathematical equation that is the model for the system you are to investigate and which you will use when you analyse your data.

When you have read the briefing, you will start planning your practical work. There is no research time, so you will need to have carried out a lot of practical work before you start; you also need to know which graph is used for each type of equation. It is likely that the assessment will take place in the second term of Year 13.

As soon as you have read the briefing you will write your plan. You will have the list of assessment criteria with you. This will help you to keep in line with what you have to do to score marks. Writing the plan should take no more than about 1 hour.

Implementation and measurements

The implementation and measurements section is worth 6 marks.

Next you carry out a plan. This could be the plan that you wrote, one that your teacher wrote or a plan from Edexcel. Your teacher will tell you which you are to use.

This part is really just ordinary laboratory work (such as that which you have done during your time studying physics) and will probably take less than an hour. You have to record your results, think about how good they are and decide whether you need to change your plan in order to improve them.

Analysis

The analysis section is worth 18 marks.

You will need to plot a graph of your results so that you have a straight line and then to find its gradient. From this you will be able to make calculations to see if the mathematical model in the briefing was true for the work you did and to see how well you have met your aim. You should also think about the uncertainties in your work and finish with a conclusion that relates to the aim on the briefing sheet. You should examine how well your experiment worked, suggest improvements and think about how you might develop the work in the future.

The analysis should take about an hour.

- **(1)** Read the briefing.
- **(2)** Plan the work.
- **(3)** Carry out your work.
- **(4)** Analyse the results to draw a conclusion.
- **(5)** Evaluate your conclusion.

Summary

Assessment criteria

You will find a copy of the assessment criteria most helpful. A complete list of these is available from the Edexcel website at www.edexcel.com/quals/gce/gce08/physics/Pages/default.aspx

If you click on 'Assessment Materials' you will find a document called 'GCE Physics 2008 marking grids A2', this contains an exact statement of all the criteria. There is 1 mark for each of the criteria and they are all independent. Since the criteria are short, what follows is an indication of what you should do in order to score the mark in each case.

Planning

Ref	What you must do to score the mark
P1	In advance, write a simple list that identifies the most appropriate apparatus required for the practical. Make sure you identify everything you will need.
P2	Your list should provide clear details of the apparatus required, including approximate dimensions of items and/or component values. All the necessary detail should be included so that exactly the right apparatus can be set out for you.
P3	Draw a large, clear and appropriately labelled diagram of the apparatus, showing how it will be set up. You should include any lengths and show how these will be measured. Provide as much information and detail as you can.
P4	State clearly how you will measure one quantity using the most appropriate instrument.
P5	Explain your choice of measuring instrument by referring to the scale on the instrument as appropriate and calculating the likely percentage error introduced by the precision of the instrument. This might be quite small.
P6	As for **P4**, state clearly how you will measure another quantity using the most appropriate instrument. This can be any measurement that you intend to make, possibly even using the same instrument to make a different measurement.
P7	This is the same as **P5** and is also a development from AS, so you will need to be quite detailed.
P8	Demonstrate how you will use your apparatus to improve both precision and accuracy by using correct measuring techniques. Describe what you will do as part of your method **P16**.
P9	Make sure that your experiment will be a fair test by identifying all other relevant quantities and explain how you will control them.
P10	It may be good practice to repeat readings. You should comment on whether repeating readings is appropriate for this experiment.
P11	Consider the safety aspects of your work and comment on all relevant points. Having identified a hazard, it is possible that no precaution is needed. In this case, you should explain why; it might be because the risk is small.
P12	Decide which graph to plot to show the data collected. Discuss how the data will be used to find what is called for in the briefing. You must be specific.
P13	Your method and instruments might introduce uncertainty and/or systematic error in your measurements. Identify the sources of uncertainty/error, so that you can take appropriate steps to reduce these.

Ref	What you must do to score the mark
P14	Your plan should contain few grammatical or spelling errors, particularly of technical terms.
P15	Your plan should be structured, using appropriate subheadings.
P16	Your method and overall plan should be clear on first reading. The method does not have to be at the end — it would be better to place it earlier.
	Maximum marks for this section — 16

Implementation and measurements

Ref	What you must do to score the mark
M1	Record all measurements in a table, with the precision shown in your plan.
M2	The precision of the instruments will introduce some uncertainty in your measurements. This should be recorded in your table.
M3	Make sure you use the correct units for your measurements.
M4	Refer to your initial plan and modify it if you think you can improve it. If you decide that no change is necessary, you should explain why you think this.
M5	You have to obtain at least six measurements for your graph.
M6	Make sure your measurements cover an appropriate range of values. Where possible, this is achieved by doubling or halving the initial value.
	Maximum marks for this section — 6

Analysis

Ref	What you must do to score the mark
A1	Your graph should have axes labelled clearly and correctly (including units).
A2	Choose scales that ensure that your points occupy as much of the paper as possible, and are easy to read.
A3	The points should be accurate to the nearest millimetre.
A4	The line of best fit may be either a straight line or a smooth curve; it should have points both above and below it.
A5	Use your graph to find the relationship between the two variables or determine the constant, as asked for in the briefing.
A6	Plot a graph that you expect will give a straight line, possibly using logarithmic axes.
A7	Make careful and accurate measurements to determine the gradient. Use a large triangle covering one-half of the plotted distance on both axes.
A8	The value of the gradient will be related to the units on the axes; both should be used to evaluate your conclusion.
A9	Use an appropriate number of significant figures (usually 3) throughout.
A10	Describe how the relevant physics principles apply to your experiment.
A11	Use with care the terms precision and either accuracy or sensitivity. These will be appropriate in your plan and your analysis.
A12	Consider the uncertainties in your results by considering how well the work has gone. Look at how close repeat readings are and whether the points are close to the best-fit line.
A13	Calculate the percentage uncertainties in your measurements.

Ref	What you must do to score the mark
A14	Based on **A13**, derive the uncertainty in your final result.
A15	Consider how realistic modifications to reduce error and improve the experiment could be introduced.
A16	State a clear, valid conclusion about the final value with its uncertainty and with a comment relating it to the briefing.
A17	Evaluate your conclusion by considering the data and the aim of experiment.
A18	Suggest further work to develop the aim.
	Maximum marks for this section — 18

Summary

- The tables on pp. 38–40 lay out what you need to do to get each mark. However, you have to apply the advice to the particular practical you are carrying out. What is given here is only the bare bones.

Development from AS

Written work

There is no visit or case study, so the 11 marks available for these at AS are given for A2 skills in the three areas of planning, implementation and measurements, and analysis. The main development is that the analysis is worth 18 marks. This is because the relationship between the variables is non-linear. This means that there might be a power relationship. An example is the relationship between velocity, v, and height, h, for an object dropped from rest and falling freely under gravity — $v^2 = 2 \times g \times h$. Here the power is 2, because the variables are h and v, and v is squared. Other more complex formulae are likely to contain exponential terms and these require logarithmic graphs. So graph plotting at A2 is more demanding than at AS.

The treatment of uncertainties is also more demanding. Here, they must be combined, or compounded, according to the relationship between the variables. This is in addition to finding percentage uncertainties in repeated readings. Uncertainty is a theme that will be looked at more closely because you will use uncertainties in evaluating the success of your work.

Examiner tip

Make sure you are familiar with which graph to plot for each relationship. Remember that you will need to know this before you start planning because you cannot look it up.

Practical work

Where possible the practical work is based on Units 4 and 5 since these have the appropriate level of demand for A2. Two topics introduced now are **oscillations** and **capacitors**. The analysis of data and graph plotting are major developments and skills you will need to practise. Here are some examples:

A simple **oscillator** such as the simple pendulum is an example of practical work with a theory demand higher than AS, even though the work is straightforward.

The period, T, is related to the length, l, by the relationship:

$$T = 2\pi \sqrt{\frac{l}{g}}$$

So the data recorded are period T and length l and the graph plotted is T^2 on the y-axis and l on the x-axis, giving a gradient of $4\pi^2/g$.

Other oscillators might be a compound pendulum or a mass on a spring.

Capacitors are devices that store charge. When a charged capacitor is discharged through a resistor, R, from an initial potential difference V_0 then the potential difference, V, across the capacitor varies with time, t, according to the relationship:

$$V = V_0 \exp(-t/RC)$$

where C is the value of the capacitance. In this case a logarithmic graph is needed and the relationship becomes:

$$\ln V = \ln V_0 - t/RC$$

Therefore, a graph of $\ln V$ on the y-axis against t on the x-axis is a straight line whose gradient is negative and $1/RC$ in value.

You have to display your data as a straight-line graph and this will require some manipulation of the formula. When you start to do this, you should:

- write the form of the equation you will use to plot the graph
- write what you will plot on the two axes
- write what the gradient and intercept will represent
- state how you will find a value for the quantity under investigation

All the above is needed for **P12**.

Some examples are given in the following table.

Relation	Equation to plot	y-axis	x-axis	Gradient	Intercept	How to find value
Electrical resistance	$V = IR$	V	I	R	0	Mean resistance is gradient
Time period of simple pendulum	$T = 2\pi\sqrt{\dfrac{l}{g}}$	T^2	l	$4\pi^2/g$	0	g is $4\pi^2/$gradient
Time constant for capacitor decay	$V = V_0\exp(-t/RC)$	$\ln V$	t	$-1/RC$	$\ln V_0$	Time constant is $-1/$gradient
Time period for linked oscillators	$T = kx^n$	$\ln T$	$\ln x$	n	$\ln k$	n is the gradient and has no units

Knowledge check 13

Light intensity varies as an inverse square with distance. When measured with an LDR, the resistance R increases with distance d. $R = kd^p$ where k and p are positive constants. What graph would you plot to display the data as a straight line and how would you find a value for p?

Themes

There are four themes that run through the three stages of the practical work. Linking together all the marking points in a theme will make it easier for you to score a high mark. The themes are uncertainties, techniques, measurements and communication.

Uncertainties

You are expected to consider how strong your final conclusion might be and to help this you should take into account the uncertainty in your work; these considerations are expected to be numerical. Eight of the criteria require you to consider the uncertainties directly. The obvious ones are **P13**, **A12**, **A13**, **A14** and also **M2** because it refers to your table of results. You should be thinking about the uncertainty in your measurements when choosing your measuring instruments, so **P5** and **P7** are also relevant. You should use uncertainties when discussing your conclusion for **A17** and this will be based on your readings and your graph.

Planning

For **P5** and **P7** you need to explain your choice of measuring instrument. You will notice that these criteria are the same as those for **P3** and **P5** at AS, so in order to demonstrate some progression from AS you will need to say more for Unit 6 than you did for Unit 3. The easiest way to do this is by considering the precision of the instrument and then calculating the likely uncertainty that this alone will introduce into your measurements.

The precision of a metre rule is 1 mm. If you are measuring a voltage using a digital multimeter on the 2 V scale then the last digit on the display shows 0.001 V or 1 mV, so the precision of the multimeter as a voltmeter *on this range* is 1 mV.

You now use this precision to calculate the likely uncertainty in your measurements. You need not find a percentage uncertainty for each reading, but it is worth considering a typical value. If you measure a range of lengths from 40 cm to 1 m using a metre rule then the uncertainty is at least 1 mm, so the likely percentage uncertainty is (1/700) × 100 (just over 0.1%), which is very low. The calculation is in millimetres and is divided by 700 mm because this is the middle of the range of readings.

When recording voltages between 0.0 V and 1.5 V, the percentage uncertainty of 1 mV becomes (0.001/0.75) × 100 (just over 0.1%). This very low uncertainty suggests that the voltmeter on this range gives readings that have very small uncertainty due to the instrument. The division by 0.75 is because this is the middle of the range of readings.

For P13 you are asked to look ahead and consider how your method might introduce further uncertainties into your measurements. If your work requires you, say, to measure the temperature of a thermistor then stabilising the temperature of the thermistor and making it the same as the thermometer will take some doing. You will have described how you did this for **P8**, but you should include it as a factor in your overall considerations for uncertainty.

The **precision** of an instrument is the smallest measuring division of that instrument.

Measuring

Measurements should be as accurate as possible.

Accuracy depends on both the method used and the skill-level using the instruments. It does not depend on the precision of the instruments. You should quantify your estimate of the uncertainty (*U*) in your measurements before you start to take them and for **M2** you should include that estimate in the table of results. The easiest way is to include a ± *U* at the foot of each column or in the header.

When recording a voltage between 0.0 V and 1.5 V the precision of the voltmeter is 1 mV (see p. 43). However, if you find that the last digit wobbles about and won't stay still then you might record an uncertainty of 5 mV. You should show this in your table of results. Each column should include a heading, such as *V*/V, with ±0.005 V alongside or at the bottom of the column.

In a capacitor discharge, it is usual to take a reading every 10 seconds. Recording the times as *t*/s = 10, 20, 30, 40... suggests that your measurement of time is to the nearest 10 s! This is probably not what you mean. You must record the time as *t*/s = 10.0, 20.0, 30.0... to show you are measuring the time to a precision of 0.1 s, which is probably what you said in your plan.

Analysis

When evaluating your results for **A17**, you should consider the uncertainty of the work and the strength of your conclusion. Think about anomalous readings and also the straightness of your line of best fit.

For **A12**, you should look back at your results and think about how easy (or not) they were to obtain. You should consider what you said for **P13**, which was a look ahead, For example, suppose your experiment involved measuring the resistance of a wire as the temperature changed. For **P13**, you will probably have said that it will be important to ensure that the thermometer and the wire are at the same temperature when you measure the resistance. How quickly the temperature is changing is a factor here. For **A12**, you can judge how successful you were in slowing down the rate of rise in temperature. You can also look at your graph; if the points lie on a good straight line you can say that there was little effect from the uncertainty of your method.

For **A13** you should calculate the percentage uncertainties in both variables and in any other measurement made. You should have done this for Unit 3. The actual uncertainty can be taken as the precision of the instrument (at least) and must be based on readings; if you took repeat readings then use either the range or half the range.

For **A14**, you are expected to combine the uncertainties, which is a new skill for A2. This can be done in two ways — either by adding fractional uncertainties or by using error bars on your graph.

Combining uncertainties

Adding fractional uncertainties

Suppose you are finding the resistivity of a metal using a wire sample.

The equation for resistivity is

$$\rho = \frac{RA}{l}$$

where R is the resistance of a length l of a wire with cross-sectional area A.

Cross-sectional area is calculated from the measurement of the diameter d as $A = \pi d^2/4$. So you have:

$$\rho = \frac{R\pi d^2}{4l}$$

You make the following measurements, with the uncertainties shown:

$$R = 31.0 \pm 0.5\,\Omega \qquad d = 0.190 \pm 0.002\,\text{mm} \qquad l = 790 \pm 2\,\text{mm}$$

This gives:

$$\rho = \frac{31.0 \times \pi \times (1.90 \times 10^{-4})^2}{4 \times 0.790} = 1.11 \times 10^{-6}\,\Omega\,\text{m}$$

The uncertainty in this value is given by:

$$\delta\rho/\rho = \delta R/R + 2\delta d/d + \delta l/l$$

$$\frac{\delta\rho}{\rho} = \frac{0.5}{31.0} + 2 \times \frac{0.002}{0.190} + \frac{2}{790} = 0.016 + 0.021 + 0.003 = 0.0400$$

So the percentage uncertainty is given by $0.040 \times 100 = 4\%$.

Note that:

- the fractional uncertainties are calculated by dividing the actual uncertainty by the mean value measured in *the same units*
- the fractional uncertainty for the diameter is doubled since the diameter is raised to the power of 2 in the original equation (it is d^2)
- the fractional uncertainty for length is added, it does not matter whether the term in the equation is the numerator or divisor; it has not been raised to a higher power

Using error bars

When the relationship between the variables involves an exponential term or an unknown power it is usual to plot the graph of the results using logarithms. This makes the treatment of the uncertainties more complex and it is much easier to use error bars. To score the mark for **A14**, using error bars is a two-stage process. First the error bars must be added to the graph and then they must be used to find the uncertainty in the gradient.

Suppose that you are plotting a graph for the discharge of a capacitor through a resistor. The potential difference and time are measured and then a graph of $\ln V$ against t is plotted (see Figure 9). The size of the error bars comes from the uncertainty in the measurement or from the spread of the readings when they were repeated.

> **Examiner tip**
> The coursework moderator expects to see treatment of the uncertainties giving an estimate. A full statistical approach is certainly not needed.

> **Examiner tip**
> It is a common mistake to draw the error bars and do nothing else. You should draw two lines for two values of the gradient; the lines should be the steepest and least steep lines that just pass through the error bars. Remember the aim of the exercise is to find an uncertainty in your final result.

There is an uncertainty in the time of 0.1 s, since the stopwatch is hand-operated, and suppose there is an uncertainty in the voltage of ±0.8 V (this is large to make the uncertainty calculation easier to understand). Consider the point plotted at $t = 22$ s and $\ln(V/V) = 2.262$, which corresponds to a potential difference of 9.6 V. We choose this point because it is a typical point. The data are plotted as a cross and we now think about error bars. Adding and subtracting the uncertainty in the potential difference gives the values shown in the table. Now calculate the natural log of these two extreme values as shown:

V/V	8.8	9.6	10.2
ln (V/V)	2.18	2.26	2.33

The difference between these two extreme values and the original value for the log gives you the length of the error bar. Here the error bar stretches 0.075 up and down and this is three divisions on the graph shown in Figure 9. The error bar stretches from 2.18 to 2.33.

The error bar on the time axis is too small to plot. You can make all the error bars the same length to avoid doing too many calculations — remember we are after an estimate of the uncertainty in the final result. Figure 9 shows the graph for the capacitor discharge with the error bars for each point.

Two lines are added to the graph — the line that is steepest and just passes through all the error boxes and the least steep line that also passes through all the error boxes. The mean value of the gradients of these two lines gives the gradient for the equation; the difference between the gradients of these two lines gives the uncertainty in the gradient.

Here is the calculation for the gradient of the graph in Figure 9.

To make it simpler, make the triangles from the $\ln V$ axis down to the line $\ln V = 1$. This gives values that are easy to read.

Steepest line: $\dfrac{2.79 - 1.00}{0.0 - 78.3} = \dfrac{1.79}{-78.3} = -0.0229$

Least steep line: $\dfrac{2.69 - 1.00}{0.0 - 81.7} = \dfrac{1.69}{-81.7} = -0.0207$

The mean value for the gradient is −0.0218 with an uncertainty of ±0.0011, which is 5.0%. In this way the error bars have been used to find the uncertainty in the gradient.

Note that when used in calculations the gradient has units of s^{-1}.

Conclusion

For **A17** you have to try to justify your conclusion. The best way to do this is with reference to your data and uncertainties. Much will depend on the briefing; you might have a definite value to compare against the value you found. You will always have your uncertainties and you can also describe how well the data fit the line of best fit.

Look again at Figure 9. The potential difference falls as the capacitor discharges. The time constant τ for the discharge is given by $\tau = -1/\text{gradient}$. In this case, $\tau = -1/(-0.0218) = 45.9$ s.

The percentage uncertainty in the time constant is the same as the percentage uncertainty in the gradient — that is one of the good things about using percentage uncertainties. So, $\tau = 45.9\,\text{s} \pm 5.0\%$.

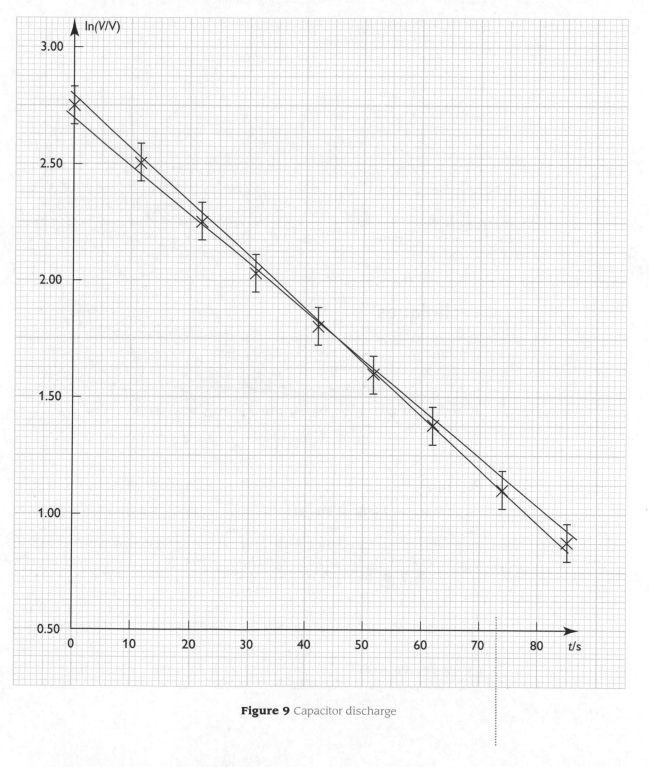

Figure 9 Capacitor discharge

If you were evaluating measurement of this time constant you could be confident that the value is accurate (close to the true value) because the points all lie close to the straight line predicted by the mathematical model and the uncertainty is small enough — you might take 10% as the upper limit.

You do not have to use error bars in deriving the uncertainty in your final value, but they are particularly useful if you have plotted a log graph. In evaluating your work the uncertainty in your final value is most important.

Summary

- Uncertainties are a feature of your work throughout the whole investigation and using them is worth 20% of the marks.
- When choosing your instruments, you need to consider their uncertainty and the effect this will have on your measurements.
- To evaluate the uncertainty in your method, you must consider the range of your repeated readings
- You must combine the uncertainties to produce an overall uncertainty in your final answer and then use this in evaluating your work.

Techniques

What you actually do with the apparatus in a practical is of prime importance. Therefore, in an assessment you must say carefully what you plan to do and say why you will do it this way. You must analyse how successful your techniques were and for **A15** suggest an improvement. The assessment criteria in this central theme are:

P1, **P2**, **P3**, **P4**, **P5**, **P6**, **P7**, **P8**, **P9**, **P10**, **P11**, **P13**, **P16**, **M4**, **A12**, **A15**, **A18**.

These cover the apparatus you use, how you use it and your evaluation after you have used it.

Planning

For **P1** you must include all the apparatus you need to do the practical work. For **P2** you must include the extra details; these might be the mass of the pendulum bob, the length of the piece of string you will need to use or the ranges of the meters. The idea is that someone else could use your list to get on with the work described in the briefing with no further help. **P3** is for a diagram that shows how this apparatus fits together. It should include physical dimensions and show how a ruler, if you need one, will be used to take the readings. This is best done with dimension lines that show, for example, that the length of the pendulum is measured to the centre of mass of the bob.

P1, **P2** and **P3** provide a comprehensive guide to what you will need to do the practical work. Figure 10 shows how you might measure the extension of a spring when a weight is hung from it.

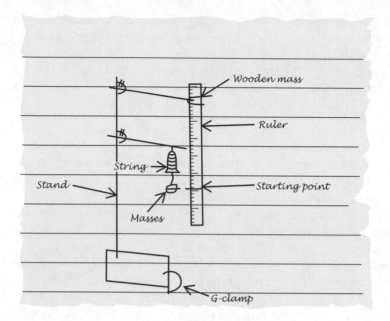

Figure 10

Compare this with the textbook diagram shown in Figure 11.

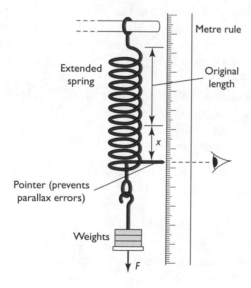

Figure 11

Examiner tip
Although you cannot produce diagrams as elegant as Figure 11, you could practise by copying this one. Try now — you should be able to produce a good copy in less than 2 minutes.

P4 to **P7** are the same as **P2** to **P5** in Unit 3 and are about choosing apparatus.

P4 and **P6** are awarded for simply stating that you will use X (an instrument) to measure Y (a variable). So, for example, you could say 'I will use a metre rule to measure the length of the string'. The wording is the same as for **P2** and **P4** at AS and you need do only the same as you did then.

P5 and **P7** are worded the same as **P3** and **P5** at AS, but you need to do more here than you did at AS in order to show progression. To justify your choice of

Knowledge check 16

You choose to measure a potential difference of about 0.7 V using a digital multimeter on its 2 V scale. The least significant digit is 0.001 V. Show that the uncertainty introduced into this experiment by this instrument is about 0.1%.

apparatus you should calculate the likely percentage uncertainty caused by using the instrument you have chosen. If you have selected a metre rule to measure the length of the string and you expect a middle length to be about 350 mm, then you divide the precision of the instrument by the middle value and multiply by 100. So the percentage uncertainty introduced by the metre rule in this measurement is $100 \times (1\,\text{mm}/350\,\text{mm})$ or $\%U = 0.3\%$.

Now that you have chosen your apparatus this is a good moment to write out your method. This is best done using bullet points. It need not be very long but should include each step. This will score **P16** and help you with the other planning criteria.

- **P8** is awarded for using good measuring techniques. These are anything that improves the accuracy or reliability of the readings.
- **P9** is for ensuring that you only change the variable you are measuring. You probably did this for GCSE and it is no more complicated here.
- **P10** Repeat readings are a good idea as they help to eliminate anomalies and you can improve accuracy by finding a méan of your repeats. Some methods do not let you repeat individual readings, in which case you should take more readings than normal — this is true when you are measuring something against the clock. Repeating the whole method is also a good idea in this case. However, you cannot take averages since you are doing a different experiment. You can show all your data by drawing two lines on your graph.
- **P11** Safety is not usually an issue in an A-level physics laboratory, but you must say why you think the experiment is safe. Usually the masses and energies are too small to hurt or the potential differences too small to shock, but a wire under tension does require you to wear goggles. You must give a precaution for each hazard identified or an explanation of why it will not cause harm. Do not invent a hazard where none exists.
- **P13** is awarded for looking forward and trying to forecast where uncertainties will occur. These are as likely to come from your method as they are from your measurements and you should mention both sources when considering the uncertainties of your techniques.

Bullet points mean you are less likely to miss out anything but you should write about your method in at an early stage for **P16**.

Examiner tip

Write your method just after you have justified your use of instruments (**P4** to **P7**). It is better not to leave it until the end.

Measuring

M4 is for reviewing your method as you take your measurements. To do this, you should think about the method you are using and look at your results. If you are finding it easy to take the readings, then there is no reason to change your method. To score **M4** you need to say why you are not changing your method — the explanation above provides a reason. If you think of a way to make the readings better then you can try it out, you do not have to stick rigidly to your plan.

Examiner tip

You must explain why you are, or why you are not, changing your method.

Analysis

You will only be awarded **A12** if you refer to your results while discussing the uncertainties in your experiment. In **P13** you looked forward and now, as you look back, you can judge how well it has gone. You should look at your table of results to see if the repeated readings are close together. You should look at your graph to see if the points all lie close to the best-fit line and you should consider which readings were difficult to take. **A12** must be more than a re-statement of **P13**.

A15 is for suggesting changes that will improve your method. This might be by using different apparatus or by doing something differently — the apparatus you mention here might not have been available at the start of your work. What you choose here must be practical and should address the difficulties you highlighted in **A12**. Light gates or 'a computer' are easy suggestions but you must say how these will be used. A diagram will be the best thing for that. **A18** is difficult. To score this mark you need to think how you might develop the work in the context of the briefing. You will almost certainly suggest different apparatus and you should write a brief plan of what you would do. If you have been investigating an oscillating system, you might think of the scale of the apparatus needed by the context of the briefing or it might be that damping would change the outcome. Describe how you might investigate this.

- The apparatus you choose and the way you use it are important.

- You must pay close attention to the wording of all the criteria.

Summary

Measurements

M1 is for recording your measurements with appropriate precision, i.e. the precision you stated for **P5** and **P7**. So if you said that your metre rule had a precision of 1 mm then your readings should be to 1 mm, unless your method means this is too difficult — reading the amplitude of a pendulum swing, for example. If using masses you can expect them to be accurate to the nearest gram and so record 200 g as 0.200 kg — 0.20 kg means something different.

M3 is for recording your results with appropriate units, so your table heading should be the same as a label on the axis of your graph, t/s or V/V. To score **M5** you should take enough readings to plot your graph; this is at least six for a straight-line graph. For **M6**, one of the variables should at least double. The briefing sheet might help you decide on your range.

M4 prompts you to consider your measurements *as you are taking them*. If you are getting the sort of readings you were expecting and they are not proving too difficult to read then you have no reason to change your plan. If you find difficulty somewhere you should write down the change in your plan and give an explanation.

Examiner tip
Draw up your results table before you start taking readings. Include the instrument reading uncertainty using ± on the first line — this scores **M2**. You do not have to decide how many rows your table should have, don't draw a line across the bottom until you have finished.

Examiner tip
For **M4**, if you are making no change to your method you *must* write down why this is the case. This shows that you have considered your plan while carrying it out.

- You will score most marks by taking careful measurements and recording them exactly.
- You should include the uncertainty in your measuring instrument.

- Think what you might do differently to make your measurements more accurate, precise and reliable.

Summary

Communication

Much of this work involves describing and explaining what you intend to do or have done, so your use of English is important. So too is your use of maths as a language to help you through to a final conclusion.

Use of drawing

If you can draw a good diagram you will be able to communicate more clearly what you will do with your apparatus and this will make it easier for you to write out your method. There is a planning mark **P3** for this, but if you think of **P1, P2, P3** and **P16** together then you will get your report off to a great start. Here are some tips on how to do this well:

- Bring the right equipment in to the laboratory — two pencils, an eraser, sharpener and ruler.
- Draw a large diagram.
- Show how you will support the apparatus, if appropriate.
- If you need a ruler, show it positioned close to the measuring point.
- Use dimension lines to indicate the actual measurements — e.g. line of sight.
- You do *not* need to include a picture of apparatus that does not connect with anything else, such as a stopwatch or micrometer screw gauge.

Use of English

When writing your plan you must say clearly what you going to do. A list of bullet points is a good way to help you present your ideas in a logical order. **P14** looks at the spelling of technical terms only. However, your method must be clear to the reader and good grammar helps. **P15** is awarded if your plan is structured using subheadings.

A11 You must use the words 'precision' and 'accuracy' correctly. These occur in the plan as often as they appear in the analysis so make sure you get them right. Students are often so keen to include them that they say '...to improve precision and accuracy...' and this is almost always wrong.

You can improve accuracy by repeating readings, but not the precision. You can improve precision by measuring more than one of the things at the same time — periodic time for a pendulum — but this will not improve accuracy, since you are still only making one measurement. Use these words correctly.

A16 and **A17** Your conclusion and support for it are the whole point of the exercise. You should make it very clear what you have found and why you think it is true. You will use mathematics in support, but make sure that what you write about is in response to the demands of the briefing sheet.

Examiner tip

Use the criteria as your subheadings, not necessarily in exactly the same order as on the marking grid. You can keep a check on your progress by doing this.

Examiner tip

Write your conclusion as a separate statement starting on a new line and make sure it refers to the briefing.

Use of mathematics

The choice of graph for different equations was covered on p. 42. Here, the broader use of maths as a language that makes the physics easier is considered.

Modelling

You are carrying out an investigation into a physics system and any physics system has a mathematical model to describe how it works. You will usually find this model as an equation on the briefing sheet. One of your planning tasks is to decide which graph to plot to ensure that your data points lie on a straight line. For **A6** you must plot the graph correctly; for **A10** you should write something about the background to the model and describe how the laws of physics apply to your apparatus. This might be to do with simple harmonic motion (shm) or with current flow as a capacitor discharges. You can also derive the logarithmic version of your equation and explain how that gives you a straight line on your graph. In your conclusion for **A17** you might refer to the model and see if it has any limitations as shown by your results. You can address these in developing the work for **A18**.

Graph plotting

A1 to **A8** are for plotting your graph and using it to derive mathematical conclusions.

A6 is for plotting correctly the graph you specified in **P12**. **A1** is for having the correct axes, as described in your plan, and for including the correct units. The gradient of a graph has no units because the axes are labelled with units — thus you plot pure numbers and hence there are no gradient units. So if you are plotting a length, l, measured in millimetres you should label the axis l/mm. If plotting the time period squared, measured in seconds squared, the label is t^2/s^2. If you are plotting a logarithmic graph with time in seconds on one axis, the label is $\log(t/s)$, or if time squared the label is $\log(t^2/s^2)$.

A2 One purpose of a graph is to display data. The data points should occupy at least half of both the axes of your graph; you should not include the origin unless you can do so without confining the points to one corner. The other purpose of a graph is to enable you to take readings from intermediate points (interpolate), so make sure your scale is sensible and easy to read, usually 1, 2, 4 or 5 units per cm of graph. In **A3** plotting should be accurate to the nearest millimetre.

For **A4** the best-fit line should have points above it and points below it and should not necessarily pass through the origin. If you are plotting a current–voltage graph then you should think of the origin as another data point. The graph might not be straight over the whole data range, in which case draw a straight line as far as it fits and subsequently draw a curve. It may be that your mathematical model does not fit the data and the points really do follow a curve. If the top and bottom points are below the line and the middle points are above the line then the graph should be drawn as a curve.

Examiner tip

When candidates choose an awkward scale they usually make a mistake in plotting the points and lose A3 as well as A2.

A current–voltage graph has a data point at the origin because when there is no potential difference across a component there is no current flowing through it. This is (0,0) or the origin.

Look at the graph in Figure 12, which displays a data set.

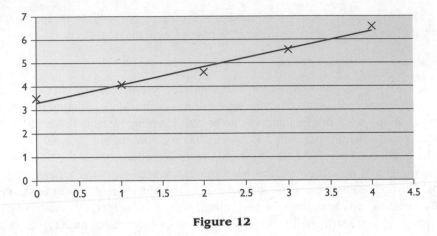

Figure 12

What is wrong? The graph shown in Figure 12 is not a good graph, for the following reasons:

- There are no labels or units on either axis, so it is impossible to tell what the data are.
- The vertical scale is too small, the points are nearly horizontal. The inclusion of the origin adds nothing to our knowledge of the system — whatever it is.
- There are only five data points.
- The first three points look as though they are in a straight line and the trend after the straight portion is upward. There are definitely two portions to the graph.
- This graph was produced using Excel and using this it is difficult to draw the appropriate trend line. Drawing graphs by hand is a skill and it is sometimes the only way to present data; it is a skill A-level physics students should have.

How can the graph be improved? First, add labels and units to the axes and then use a better scale for the y-axis. The real problem is that there are not enough data to be able to tell what is going on — no conclusions can be drawn from the data. Such a graph would fail to score **A1**, **A2** and **A4**.

What if the graph is all there is? If you are doing your assessment and there is no possibility of getting more data, or checking the data set, then you should draw a straight line through the first three points and continue it with a steeper gradient for the last two. Then extend the straight portion and draw a triangle to find the gradient of that portion of the graph. You should write in your conclusion that *over this portion* the graph is straight and then you can quote a value for the gradient.

A7 is for taking measurements from your graph in order to calculate a gradient. You do this by drawing a triangle and measuring the horizontal and vertical sides. Make the triangle as large as possible; if it extends to the edge of the grid then it is easier for you to take one of the measurements. Remember to include a 'minus sign' if the gradient is negative, i.e. the line slopes downwards.

A5 is for linking the gradient to the constant mentioned in the briefing paper.

A8 is for linking the correct units to your calculated constant. The powers of ten will depend on the scale of the graph, so there can be a lot to get right here.

Examiner tip
The gradient of your graph is one of the most important parts of your conclusion. Make sure it is correct by writing down each measurement on the triangle and writing the calculation on the graph paper itself. This is another reason why the scale should be easy to read.

Edexcel AS/A2 Physics

A9 is for using the correct number of significant figures (s.f.) throughout your work. Your measurements will probably be to 3 s.f. and your graph should be plotted using 3 s.f., although there are occasions when it can be different (see below). The gradient and final value of the constant should also be to 3 s.f. Your conclusion cannot be more precise than your measurements.

If you are plotting a logarithmic graph and the first significant figure is the same each time, then plot the graph using 4 s.f.

If your readings start at 0.92 and go up to 1.56 then you can only record what your instruments read and so 2 *and* 3 s.f. is appropriate here (see **M1**).

Summary

- The use of the languages of English and mathematics can make it much easier for you to express yourself clearly.

- A drawing can communicate a great deal about your plan.

- Remember the mathematics is there to make the physics easier.

Worked examples

Choosing the graph to plot

Q **When a pendulum swings, energy is lost through damping and the amplitude decreases. After a number of swings, n, the amplitude, A, is given by $A = A_0 \exp(-kn)$ where k is the damping constant and A_0 is the initial amplitude. You record values for A and n. What graph would you use to display your data as a straight line? How would you find a value for k?**

A Take logs of the equation to give $\ln A = \ln A_0 - kn$.

You should plot $\ln A$ against n which will be in the form:

$$y = mx + c$$

$$\text{since } \ln A = -kn + \ln A_0$$

If this is a good mathematical model for the pendulum the result will be a straight line with a negative gradient equal to the damping constant. The theory of shm only holds true if the amplitude is small.

The following question is similar to the previous one.

Q **The current, I, through a diode with a constant potential difference across it will vary with temperature, T, according to the model:**

$$I = I_0 \exp(-q/T)$$

where q is a constant and I_0 is the current at zero temperature.

What graph would you use to display your data as a straight line? How would you find a value for q?

A Take logs of the equation to give $\ln I = \ln I_0 - q/T$.

So you should plot $\ln I$ against $1/T$ which will be in the form:

$$y = mx + c$$

$$\text{since } \ln I = -q/T + \ln I_0$$

If this equation is a good mathematical model for the diode the result will be a straight line with a negative gradient equal to the constant, q.

The next question goes a little further.

Q An object falling freely under gravity loses gravitational potential energy, mgh, and gains kinetic energy, $\frac{1}{2}mv^2$. To check this relationship experimentally, you drop an object from rest and measure its velocity, v, after it has fallen from a height, h.

(a) Explain why it is not important to know the mass of the falling object.

(b) Explain how you would measure the velocity after it has fallen from a height, h.

(c) What graph would you plot to find the exact value for the power of v in the derived equation:

$$gh = \frac{1}{2}v^n$$

A

(a) The value of m is the same in both expressions, so when they are made equal they cancel out.

(b) You could arrange for the object to pass through a light gate (it could fall down a Perspex tube). Measure the length, l, of the object and find the time, t, that it takes to fall through the light gate. The velocity $v = l/t$.

(c) $2gh = v^n$

Take logs of both sides to give $\ln 2gh = n\ln v$.

This is an occasion when it is better to plot the independent variable on the y-axis.

Rearranging gives:

$\ln 2g + \ln h = n\ln v$

which becomes:

$\ln h = n\ln v - \ln 2g$

This has the form:

$y = mx + c$

So if a power law applies the graph will be a straight line. The gradient will be n and, if the initial equation is the right model, should be equal to 2. The intercept will be $-\ln 2g$, so the value of acceleration due to gravity can also be checked. Compare the value obtained for g with $9.81\,\mathrm{m\,s^{-2}}$. If it is close — say within 5% — then, if the points are close to the best-fit straight line on the graph you will be able to state with confidence that $n = 2$.

Plotting the graph

Try plotting the graphs yourself. Measure the gradients and intercepts and compare your answers with those quoted in these examples.

Q A magnet is suspended at the centre of a coil of wire. The coil is vertical and current is passed through it. The magnet is made to oscillate and the period of oscillation, _T_, is measured as the current, _I_, is varied. It is thought that $T^{-2} = kI + b$ where _k_ and _b_ are constants. Plot a suitable graph and find values for _k_ and _b_. Comment on the reliability of your values.

_I_/A	0.00	1.00	2.00	3.00	4.02	5.01
Mean _T_/s	1.23	0.827	0.673	0.581	0.52	0.475

A If you plot a graph of T^{-2} on the _y_-axis and _I_ on the _x_-axis (Figure 13) you will get a straight line with a gradient of 0.750 and an intercept of 0.693. So $k = 0.750\,\text{s}^{-2}\,\text{A}^{-1}$ and $b = 0.693\,\text{s}^{-2}$. Most of the points are close to the line but the lowest point is below the best-fit line, as is the top point. If the data extended further we might find that a curve is a better fit. In this case, the mathematical model (equation) would have to be changed. However, you can say that _over the range of the readings_ the graph is straight enough and so the values are reliable.

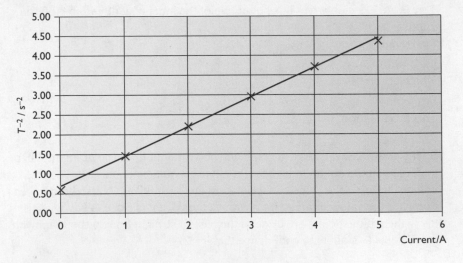

Figure 13

Q When gamma rays are passed through lead and detected with a Geiger counter the count rate, *A*, decreases as the thickness, *x*, of the lead increases according to the formula:

$$A = A_0 \exp(-\mu x)$$

where A_0 is the count rate with no lead between source and detector and μ is the absorption coefficient.

Plot the data below on a suitable graph and find a value for μ.

x/mm	0.00	6.30	12.74	19.04	25.44
A/min^{-1}	1002	739	553	394	304

A Take logs of the equation $A = A_0 \exp(-\mu x)$ to give $\ln A = \ln A_0 - \mu x$. Then:

$$\ln A = -\mu x + \ln A_0$$

which has the same form as $y = mx + c$.

If you plot a graph of $\ln A$ on the *y*-axis and *x* on the *x*-axis (Figure 14) you will get a straight line with gradient −0.0510 (or 51.0 if you convert *x* to metres) and an intercept of 6.87. Thus the absorption coefficient μ has a value of 51.0 m^{-1}.

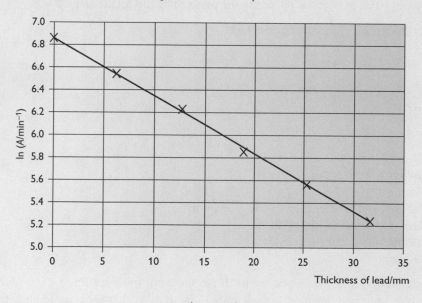

Figure 14

e There is a lot to take in here and you have done well to get through it. The advice contained here must now be applied to actual experiments. The exemplar piece of work that follows was carried out by a student.

Unit 6 exemplar

The work starts with a **briefing sheet** given to you by your teacher, which sets the work in context and tells you the end point you will be aiming for. This is usually either the value of a constant in an equation or the verification of the formula (or mathematical model) governing a particular physics system. The following is some work carried out by a student. After each section there is a commentary (preceded by the icon **e**), which explains how the marks were scored.

Briefing

The student was given the following short statement as the briefing.

You are to find the spring constant for a spring by two methods and then use both methods to find the unknown mass of another object. By comparing the results from the two methods you will be able to evaluate the reliability of your results.

The first method is by hanging masses on the spring and measuring the extension. Since we expect the force F and extension x to be related by the relationship $F = k \times x$, a graph of force against extension should be a straight line.

In the second method a series of masses are hung on a spring and made to oscillate vertically. The period T is related to the mass m by the relationship $T = 2\pi \sqrt{\dfrac{m}{k}}$.

In both cases k is the spring constant.

The student then went on to produce this report.

Planning

Apparatus

- Stand (>1 m)
- Two bosses and clamps
- Spring
- Masses and hanger (50 g to 350 g, 50 g each)
- Stopwatch (accurate to 0.01 s)
- Metre rule (accurate to 0.001 m)
- G-clamp

I will use the metre rule (see Figure 15) to measure the position of the bottom of the mass hanger; because it is accurate to 0.001 m, which is accurate enough for this experiment.

The stopwatch is accurate to 0.01 s but since my reaction time is about 0.1 s the stopwatch is more accurate than my method. Since the time I am measuring will be about 10 s the uncertainty in my timing will be about (0.1/10) × 100% = 1%, which is quite low.

Examiner tip
The terms 'accuracy' and 'precision' must be used correctly throughout the report. They are not correct here, so the candidate cannot score **All**.

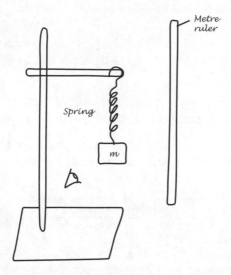

Figure 15

Method

First, I will set up the apparatus as shown in the diagram (see Figure 15). Then I will hang masses on the spring and measure the position of the bottom of the mass hanger, as shown on the diagram. This is the first method.

For the second method I will hang the same masses on the spring starting with 50 g and pull them down and let them go. They will bounce up and down. I will use the stopwatch to measure the period. I will do this by measuring how long it takes to do 20 complete bounces and then divide this time by 20. I will do this three times (five times if we have long enough) and find the mean.

There is a starting point on the rule, I will start the stopwatch when it first passes and stop it after 20. To make it a fair test I should do the experiment on the same table and with the windows shut, therefore they won't influence the experiment.

Repeating readings makes the experiment more accurate and reduces random error. A reading that is different from the others can be left out.

Safety

- I will need to give an appropriate force to the spring each time so that it won't fall off and hurt people.
- Using a G-clamp makes the stand more stable so that it won't fall over.

Graph

$T = 2\pi \sqrt{\dfrac{m}{k}}$, so $T^2 = 4\pi^2 \left(\dfrac{m}{k}\right)$.

Rearranging gives $m = \dfrac{T^2 k}{4\pi^2}$

This gives the relationship between m and T. Next I will do the experiment and take readings. By plotting the readings on a graph I will be able to find the gradient and test the relationship between m and T. Then I can find the mass of the other object.

Main sources of uncertainty

As the masses are to a precision of 0.1 g the percentage uncertainty in mass will be about (0.1/100) × 100% = 0.1%, which is very small and so will not influence the experiment.

The main source of uncertainty is from the reading of the time. Because of human reaction time there will be about 0.1 s difference in the readings.

There will also be errors in putting force on the spring. If I don't stretch the spring vertically it will influence the accuracy of the experiment

Ref		Commentary on the planning marks
P1	✓	The list contains everything necessary to do the experiment and bullet points make it easy to read.
P2	✓	The list provides clear details of the apparatus required, including the number and size of the masses. Including the size of the stand shows a thoughtful approach, although it is not really important.
P3		The diagram contains most of the apparatus needed. The rule (not ruler) is too far away from the mass and because the student has tried to include some perspective effect the position of the eye is simply confusing. The diagram has been drawn without using a ruler and for such a simple experiment it is poor. There is no reference to a timing marker.
P4	✓	The student plans to measure the extension using a metre rule.
P5		The precision of the rule is quoted but not related to the size of the measurement. The student should have calculated the percentage uncertainty; I mm in 30 mm would be 3% and about right. Note that **A11** is lost because the precision of the instruments is incorrectly called 'accuracy'.
P6	✓	The stopwatch will be used to measure the time.
P7	✓	The precision (wrongly called accuracy) of the stopwatch is compared with human reaction time, thus justifying the use of the instrument. The percentage uncertainty calculation, based on the method, contributes to **P13**.
P8	✓	The student plans to find the time for 20 oscillations and also states that a difference method will be used to find the extension.
P9		The statement about the windows and the table is not worth anything at this level and the comment about the starting mark is too vague. The initial amplitude of the oscillation is about the only factor that might be controlled, even if it is thought that it will not affect the outcome.
P10	✓	The student identifies the benefits of taking repeats and states that the mean will be found each time. This is a bit spread out but that does not matter.
P11	✓	The first comment about safety is not worth anything because it is too vague. The second comment concerns the apparatus and this is good.
P12		The student manipulates the equation but does not state which graph will be plotted or how the gradient will be used.
P13	✓	There is the percentage uncertainty (%U) calculation from the apparatus section; the effect of the %U in the mass is also considered. The spring will hang vertically, whether or not the rule is vertical is more to the point.
P14	✓	Grammar and spelling are fine.
P15	✓	There is an obvious structure and subheadings are used.
P16	✓	The plan is clear with the method detailed.
	12	**Marks scored**

ⓔ The precision of an instrument is the smallest measuring division; the accuracy of an instrument depends on whether it is used correctly. If there are no factors to control, for **P9**, then this should be stated with a reason. Here the student might say that since the initial amplitude has no effect on the period of shm there are no factors to control — this would also go towards **A10** because of the mention of shm. This is a good plan let down by the absence of some key detail; there is also some irrelevant detail included.

Implementation and measurements

Mass/kg	Rule mark/m	Force/N
0.10	0.345	0.98
0.15	0.367	1.47
0.20	0.389	1.96
0.25	0.412	2.45
0.30	0.434	2.94
0.35	0.457	3.43
Unknown mass	0.372	

With no weight on the spring, the reading was 0.303 m.

Mass/kg	Time for 20T in seconds				T/s	T^2/s^2
	$20T_1$	$20T_2$	$20T_3$	Mean for 20T		
0.10	8.63	8.64	8.65	8.64	0.432	0.187
0.15	10.73	10.71	10.72	10.72	0.536	0.287
0.20	12.18	12.17	12.21	12.19	0.605	0.366
0.25	13.64	13.53	13.60	13.59	0.679	0.461
0.30	14.96	14.85	14.96	14.92	0.746	0.557
0.35	15.85	15.99	15.87	15.90	0.795	0.632
Unknown mass	11.01	11.02	11.02	11.02	0.551	0.304

I planned to start with 50 g but found the oscillations were too quick to count.

Ref		Commentary on the implementation and measurements marks
M1		The design of the table is good. The readings for 20T and from the ruler are to the precision expected from the plan but those for the mass are not.
M2		There is no indication of the precision of the readings; for example the timing method was planned to be 0.1 s. An estimate of the uncertainty is expected.
M3	✓	All the units are correct.
M4	✓	The planned 50 g is not practicable to measure...
M5	✓	...which still leaves 6 readings
M6	✓	The period nearly doubles and there is a good range for the mass.
	4	**Marks scored**

Examiner tip
Quoting the mass to 2 d.p. suggests that it is being measured to a precision of 10 g. In the plan the precision of the masses is given as 0.1 g. **M1** is not awarded here because of the difference.

The precision in the results table should be that quoted in the plan.

Analysis

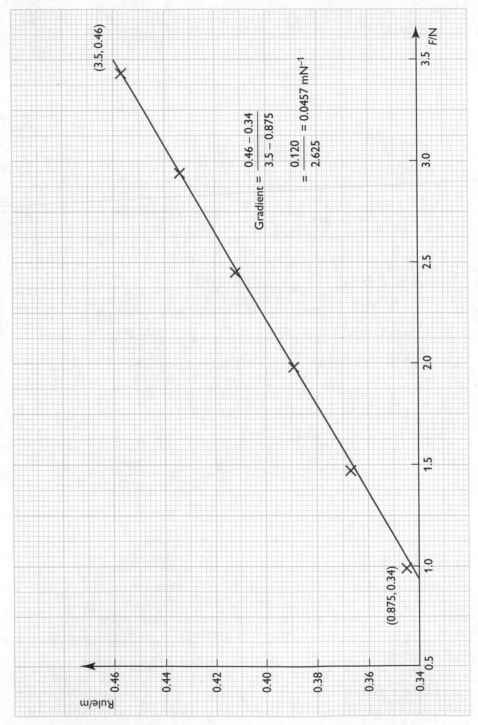

Figure 16 Graph of rule reading against force on a spring

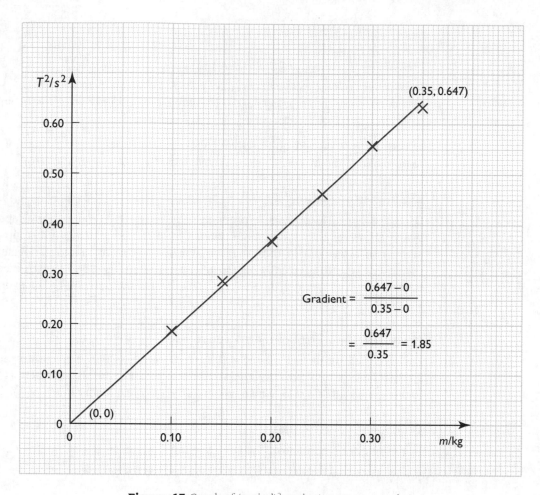

Figure 17 Graph of (period)² against mass on a spring

From the first graph of ruler reading against force (Figure 16), I find a good fit to the data with all the points lying very close to the best-fit line. Since I have not plotted extension the best-fit line will not pass through the origin.

The value for the gradient is $0.04571\,\mathrm{m\,N^{-1}}$. This gives a value for the spring constant of $1/0.04571 = 21.88\,\mathrm{N\,m^{-1}}$.

For the second graph of (time period)² against mass (Figure 17), the points all lie on the straight line of best fit and there is a zero intercept, which suggests that the variables might be directly proportional, as suggested by the shm equation, $T^2 = 4\pi^2 m/k$

Since I have plotted T^2 on the y-axis and m on the x-axis, the gradient $= 4\pi^2/k$

Since the gradient $= 4\pi^2/k = 1.85\,\mathrm{s^2\,kg^{-1}}$, this gives a value for the spring constant, k, of $21.34\,\mathrm{N\,m^{-1}}$.

Sources of error

Measuring the length of the extension

As planned the percentage uncertainty caused by the precision of the rule (±1 mm) was sufficiently small for my needs. The largest uncertainty given by the smallest length is just $(1/42) \times 100\% = 2.4\%$.

Nevertheless it proved quite difficult to accurately measure this quantity as it required lining up by eye. This could have introduced a systematic error.

Timing the oscillations

Using the fiducial marker, I have probably achieved the maximum accuracy possible by eye. However the variation in my reaction times on starting and stopping the stopwatch is evident in the range shown by my repeat readings. These seem consistent with the typically quoted uncertainty of 0.1 s. The largest percentage uncertainty is $(0.1/8.3) \times 100\% = 1.2\%$. This is also the percentage uncertainty in the period itself.

Measuring the weight of the slotted masses

I weighed all the masses using a top-pan balance, which is accurate to 0.01 g. The smallest mass was 49.95 g so the uncertainty in the masses is about $(0.05/49.95) \times 100\% = 0.1\%$ which will not affect my result very much.

Combining uncertainties

I can find the percentage uncertainties in the k values from both methods.

Spring extension: $\delta k / k = \delta F / F + \delta x / x = 0.1\% + 2.4\% = 2.5\%$

Timing: $\delta k / k = \delta m / m + 2 \times \delta T / T = 0.1\% + 2 \times 1.2\% = 2.5\%$

So it seems that the percentage uncertainty in both methods is the same.

Improvements

The largest source of error is in the timing, even when timing many cycles. This can be tackled by using an ultrasound position sensor placed under the mass hanger and connected to a data logger. This displays data as a graph on a screen and reduces the uncertainty in timing — one can also see if the period changes.

Conclusions

From a Hooke's law experiment I have found a value for $k = 21.88 \, \text{N m}^{-1}$ and from the shm experiment I have found that $k = 21.34 \, \text{N m}^{-1}$.

The mean of these two values is $21.61 \, \text{N m}^{-1}$ and they differ by $0.54 \, \text{N m}^{-1}$, so the percentage difference between the two values is $(0.54/21.61) \times 100\% = 2.5\%$. Since the percentage uncertainty is 2.5% for both experiments and the percentage difference in the k values is about the same as this, I think my mean value for k is probably accurate.

Unknown mass

From the extension graph the unknown mass made the spring 0.372 m long. From the graph this gives a value for the force of 1.575 N, which is equivalent to a mass of 1.575/9.81 = 161 g.

When oscillated on the spring the period was 0.551 s. Using the timing graph with a value of 0.304 s^2 the best-fit line gives a value of 163 g for the unknown mass.

When placed on the top-pan balance the mass was measured as 164.4 g. This is 2%, and 0.9% different from the graph values, which, again, are below the percentage uncertainty. The percentage differences are less than the percentage uncertainties, which suggests that timing a mass oscillating on a spring is a good way to find its mass.

Further work

I would use the data logger and computer to do more tests on different springs to see if they produced results as good as these.

Ref		Commentary on the analysis marks
A1	✓	The extension graph is ignored because it is too easy at this level. The timing graph has the axes labelled appropriately with the correct units.
A2	✓	The scales are easy to read and allow the points to cover the page.
A3	✓	The points checked are accurate to the nearest millimetre.
A4	✓	The best-fit line has points above it and below. It has not been forced through the origin and shows an accurate trend.
A5	✓	The value for the spring constant is derived from the gradient.
A6	✓	Although not mentioned in the plan, this is the appropriate graph.
A7	✓	The gradient measurements are accurate and the triangle uses values that are easily read at the edges of the graph. Check these on the graph.
A8	✓	The gradient calculation gives the correct value for the units.
A9		The k values are shown to 4 s.f. but the graph is only precise to 3 s.f. despite, as here, the points being exactly on ¾ divisions. Any value calculated from a graph should be to 3 s.f., as should the data plotted. The trailing zeros in gradient calculations can be ignored.
A10		Although shm is referred to it is not related to the measurements in any way. It would have been a good idea to refer to the mass–spring system and the way it moves after being displaced, accelerating towards the equilibrium point etc.
A11		The terms precision and accuracy are confused in the plan and elsewhere.
A12	✓	The uncertainties get a lot of treatment. The discussion about the length measurement is good, it might be said that the uncertainty in a difference method is twice the uncertainty of a single reading since there are two readings; so here 2 mm. For the timing the student uses an unsubstantiated figure of 0.1 s rather than using the spread in the readings — which is actually rather less. The masses used have a very small uncertainty.
A13	✓	The percentage uncertainties are calculated correctly.

Ref		Commentary on the analysis marks
A14		Although the uncertainties quoted are combined correctly — doubling that for T for example — because the uncertainties are not based on the readings taken, this mark is not awarded.
A15	✓	The data logger is described well and the student explains that this will tackle the uncertainties mentioned in **A12**.
A16	✓	The conclusion is stated clearly and the numerical values quoted.
A17	✓	The conclusion is well supported by the data gathered and the percentage difference is compared with the percentage uncertainty to very good effect. This student can feel confident that they have carried out an accurate experiment.
A18		The suggested work does little to move the experiment on. Two springs in opposition or working horizontally to remove the effect of gravity would be of interest — making that work would be a suitable challenge.
	13	**Marks scored**

ⓔ The student scored **29 marks overall** which is equivalent to the bottom of grade B. In the analysis section the student used the appropriate skills but did not relate them enough to the data recorded. The final values proved reliable and so the practical work must have been accurate and carried out well. Much of what the student has done is very good indeed. However, the extra detail as indicated in the commentaries would give this student the top grade.

Knowledge check answers

1. Resistivity ρ is given by $\rho = R \times A/l$ = (resistance × cross-sectional area)/length

   Units are $\Omega \times m^2/m = \Omega\,m$ — it is fine for your explanation (as asked for in the question) to be mathematical, as here.

2. To find the resistivity of the metal in a wire you will need: a multimeter set to read resistance, a metre rule, 1.05 m length of the wire, masking tape (to attach the wire to the rule), crocodile clips (to attach to the wire), connecting leads and a micrometer screw gauge

3.

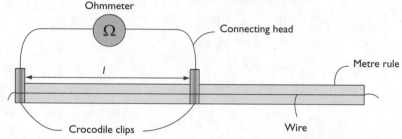

   Ohmmeter
   Connecting head
   Metre rule
   l
   Crocodile clips
   Wire

4. A micrometer screw gauge has a precision of 0.01 mm. The diameter of the wire will be about 0.25 mm, so the uncertainty will be quite small. Nothing else has such a small precision.

5. Use a timer or stopwatch.

   Set a pendulum swinging and count down to start.

   Measure 10T and divide by 10.

   Repeat your readings for 10T and find the mean.

   Did you get all three steps? Don't forget to calculate the mean.

6. Multimeter set to measure resistance in ohms. Check for zero error by connecting the probes together before attaching them to the component. You can improve the precision by selecting the range that gives the most sensitive measurement — the lowest range that gives a reading.

7. The stand might rock or fall over, so a G-clamp could be used to clamp it to the bench if this seems likely.

   The elastic rubber band might break, but the energy stored in it is quite small so this is unlikely to do any damage. However, the masses then fall so you should keep your feet out of the way. Keep an eye on the rubber band to look for signs of failure.

8. Systematic error might occur because the crocodile clips have contact resistance.

   Random error might be caused in measuring the length to the exact position of the crocodile clips, the end clip might not be at zero.

   The wire might not be of uniform cross-section along its length.

9.

Experiment	Independent variable	Dependent variable
A resistivity	Length	Resistance of wire
B Young modulus of rubber	Mass hanging	Position of bottom of mass hanger — not length or extension

10. Gradient of line of best fit
    $= \Delta R/\Delta l = (12.00 - 5.60)/(1.015 - 0.40) = 6.40/0.615 = 10.4\ \Omega\,m^{-1}$

11. The spread of readings is $4.88 - 4.65 = 0.23$. Half the range is normally used to calculate uncertainties, either 0.11 or 0.12. we shall take here 0.12.

    The percentage uncertainty can be calculated as
    $100 \times (0.12/4.75) = 2.5\%$

12. Gradient of line of worst fit
    $= \Delta R/\Delta l = (12.00 - 5.70)/(1.020 - 0.40)$
    $= 6.30/0.620 = 10.2\ \Omega\,m^{-1}$

13. Plot $\ln R$ on the y-axis against $\ln d$ on the x-axis. The graph is a straight line with a gradient equal to the constant p.

14. The mean value of $V = 4.6\,V$

    An *estimate* of the uncertainty is needed, so it is not wrong to use the whole range $\pm 0.6\,V$ or the difference between the mean and the furthest value, here 0.4 V, as the actual uncertainty. Half the range is normally used so in the example shown:

    %uncertainty $= (0.3/4.6) \times 100 = 6.5\%$. $V = 4.6\,V\ \pm 6.5\%$

15. The mean of 0.48, 0.46, 0.42, 0.45 and 0.39 is 0.44. The range is $0.48 - 0.39 = 0.09$. Take half the range as the uncertainty 0.045. So the percentage uncertainty is $100 \times (0.045/0.44) = 10\%$.

16. $100 \times (0.001/0.7) = 0.14\%$

Note: **bold** page numbers indicate defined terms.